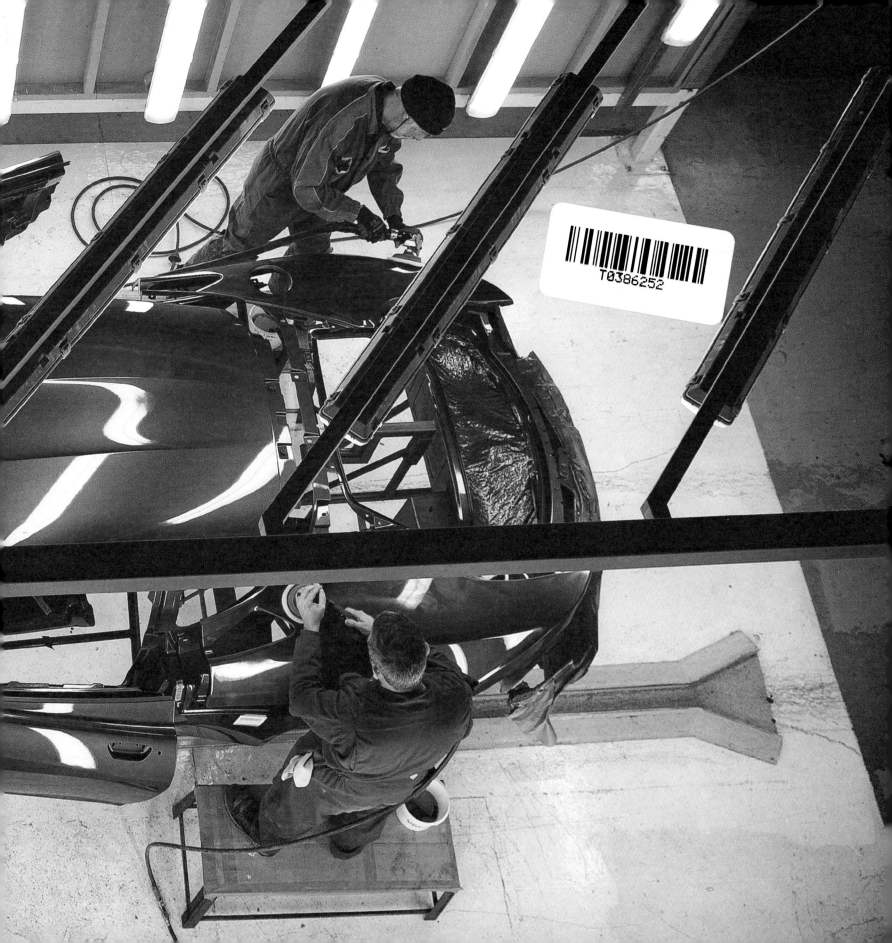

Gold Leaf Team Lotus Type 72 #3 and Lotus Evora GT410 Sport. (Courtesy Stephanie Ewen)

Sports car and motorsport titles from Veloce –
1½-litre GP Racing 1961-1965 (Whitelock)
AC Two-litre Saloons & Buckland Sportscars (Archibald)
Alfa Romeo 155/156/147 Competition Touring Cars (Collins)
Alfa Romeo Giulia Coupé GT & GTA (Tipler)
Alpine & Renault – The Development of the Revolutionary Turbo F1 Car 1968 to 1979 (Smith)
Alpine & Renault – The Sports Prototypes 1963 to 1969 (Smith)
Alpine & Renault – The Sports Prototypes 1973 to 1978 (Smith)
An Incredible Journey (Falls & Reisch)
Autodrome (Collins & Ireland)
Automotive A-Z, Lane's Dictionary of Automotive Terms (Lane)
Automotive Mascots (Kay & Springate)
Bahamas Speed Weeks, The (O'Neil)
Bluebird CN7 (Stevens)
BMC, The Cars of (Robson)
BMC Competitions Department Secrets (Turner, Chambers & Browning)
BMW 5-Series (Cranswick)
BMW Z-Cars (Taylor)
BMW Classic 5 Series 1972 to 2003 (Cranswick)
BMW M3 & M4, The complete history of these ultimate driving machines (Robson)
British at Indianapolis, The (Wagstaff)
British Cars, The Complete Catalogue of, 1895-1975 (Culshaw & Horrobin)
BRM – A Mechanic's Tale (Salmon)
BRM V16 (Ludvigsen)
Bugatti – The 8-cylinder Touring Cars 1920-34 (Price & Arbey)
Bugatti Type 40 (Price)
Bugatti 46/50 Updated Edition (Price & Arbey)
Bugatti T44 & T49 (Price & Arbey)
Bugatti 57 2nd Edition (Price)
Bugatti Type 57 Grand Prix – A Celebration (Tomlinson)
Camaro 1967-81, Cranswick on (Cranswick)
Caravan or Motorhome Habitation Check, Do Your Own (Shephard)
Caravan, Improve & Modify Your (Porter)
Caravans, The Illustrated History 1919-1959 (Jenkinson)
Caravans, The Illustrated History From 1960 (Jenkinson)
Carrera Panamericana, La (Tipler)
Car-tastrophes – 80 automotive atrocities from the past 20 years (Honest John, Fowler)
Chevrolet Corvette (Starkey)
Chrysler 300 – America's Most Powerful Car 2nd Edition (Ackerson)
Chrysler PT Cruiser (Ackerson)
Citroën DS (Bobbitt)
Classic British Car Electrical Systems (Astley)
Classic Engines, Modern Fuel: The Problems, the Solutions (Ireland)
Cobra – The Real Thing! (Legate)
Cobra, The last Shelby – My times with Carroll Shelby (Theodore)
Competition Car Aerodynamics 3rd Edition (McBeath)
Competition Car Composites A Practical Handbook (Revised 2nd Edition) (McBeath)
Cool Recipes & Camping Hacks for VW Campers (Richards)
Concept Cars, How to illustrate and design – New 2nd Edition (Dewey)
Cortina – Ford's Bestseller (Robson)
Cosworth – The Search for Power (6th edition) (Robson)
Coventry Climax Racing Engines (Hammill)
Daily Mirror 1970 World Cup Rally 40, The (Robson)
Daimler SP250 New Edition (Long)
Datsun Fairlady Roadster to 280ZX – The Z-Car Story (Long)
Dino – The V6 Ferrari (Long)
Drive on the Wild Side, A – 20 Extreme Driving Adventures From Around the World (Weaver)
Driven – An Elegy to Cars, Roads & Motorsport (Aston)
Essential Guide to Driving in Europe, The (Parish)
Fast Ladies – Female Racing Drivers 1888 to 1970 (Bouzanquet)
Ferrari 288 GTO, The Book of the (Sackey)
Ferrari 333 SP (O'Neil)
Fiat in Motorsport (Bagnall)
Fiats, Great Small (Ward)
Ford versus Ferrari – The battle for supremacy at Le Mans 1966 (Starkey)
Formula 1 - The Knowledge 2nd Edition (Hayhoe)
Formula 1 All The Races - The First 1000 (Smith)
Formula One – The Real Score? (Harvey)
Formula 5000 Motor Racing, Back then ... and back now (Lawson)
Forza Minardi! (Vigar)
France: the essential guide for car enthusiasts – 200 things for the car enthusiast to see and do (Parish)
The Good, the Mad and the Ugly ... not to mention Jeremy Clarkson (Dron)
Grand Prix Ferrari – The Years of Enzo Ferrari's Power, 1948-1980 (Pritchard)
Grand Prix Ford – DFV-powered Formula 1 Cars (Robson)
Great British Rally, The (Robson)
GT – The World's Best GT Cars 1953-73 (Dawson)
Hillclimbing & Sprinting – The Essential Manual (Short & Wilkinson)
Honda NSX (Long)
Honda S2000, The Book of The (Long)
Immortal Austin Seven (Morgan)
Inside the machine – An engineer's tale of the modern automotive industry (Twohig)
Inside the Rolls-Royce & Bentley Styling Department – 1971 to 2001 (Hull)
Intermeccanica – The Story of the Prancing Bull (McCredie & Reisner)
Jaguar - All the Cars (4th Edition) (Thorley)
Jaguar from the shop floor (Martin)
Jaguar E-type Factory and Private Competition Cars (Griffiths)
Jaguar, The Rise of (Price)
Jaguar XJ 220 – The Inside Story (Moreton)
KTM X-Bow (Pathmanathan)
Lamborghini Miura Bible, The (Sackey)
Lamborghini Murciélago, The book of the (Pathmanathan)
Lamborghini Urraco, The Book of the (Landsem)
Lancia 037 (Collins)
Lancia Delta HF Integrale (Blaettel & Wagner)
Lancia Delta Integrale (Collins)
Land Rover Design - 70 years of success (Hull)
Le Mans Panoramic (Ireland)
Lola – The Illustrated History (1957-1977) (Starkey)
Lola – All the Sports Racing & Single-seater Racing Cars 1978-1997 (Starkey)
Lola GT – The DNA of the Ford GT40 (Starkey)
Lola T70 – The Racing History & Individual Chassis Record – 4th Edition (Starkey)
Lotus 18 Colin Chapman's U-turn (Whitelock)
Lotus 49 (Oliver)
Lotus Elan and +2 Source Book (Vale)
Lotus Elite – Colin Chapman's first GT Car (Vale)
Lotus Evora – Speed and Style (Tippler)
Maserati 250F In Focus (Pritchard)
The MGC GTS Lightweights (Morys)
Mitsubishi Lancer Evo, The Road Car & WRC Story (Long)
Montlhéry, The Story of the Paris Autodrome (Boddy)
Motor Racing – Reflections of a Lost Era (Carter)
Motor Racing – The Pursuit of Victory 1930-1962 (Carter)
Motor Racing – The Pursuit of Victory 1963-1972 (Wyatt/Sears)
Motor Racing Heroes – The Stories of 100 Greats (Newman)
Motorsport In colour, 1950s (Wainwright)
N.A.R.T. – A concise history of the North American Racing Team 1957 to 1983 (O'Neil)
Nissan GT-R Supercar: Born to race (Gorodji)]
Nissan – The GTP & Group C Racecars 1984-1993 (Starkey)
Northeast American Sports Car Races 1950-1959 (O'Neil)
Nothing Runs – Misadventures in the Classic, Collectable & Exotic Car Biz (Slutsky)
Porsche, Cranswick on (Cranswick)
Porsche 356 (2nd Edition) (Long)
Porsche 356, The Ultimate Book of the (Long)
Porsche 908 (Födisch, Neßhöver, Roßbach, Schwarz & Roßbach)
Porsche 911 Carrera – The Last of the Evolution (Corlett)
Porsche 911R, RS & RSR, 4th Edition (Starkey)
Porsche 911 SC, Clusker
Porsche 911, The Book of the (Long)
Porsche 911 – The Definitive History 1963-1971 (Long)
Porsche 911 – The Definitive History 1971-1977 (Long)
Porsche 911 – The Definitive History 1977-1987 (Long)
Porsche 911 – The Definitive History 1987-1997 (Long)
Porsche 911 – The Definitive History 1997-2004 (Long)
Porsche 911 – The Definitive History 2004-2012 (Long)
Porsche 911, The Ultimate Book of the Air-cooled (Long)
Porsche – The Racing 914s (Smith)
Porsche 911SC 'Super Carrera' – The Essential Companion (Streather)
Porsche 914 & 914-6: The Definitive History of the Road & Competition Cars (Long)
Porsche 924 (Long)
The Porsche 924 Carreras – evolution to excellence (Smith)
Porsche 928 (Long)
Porsche 930 to 935: The Turbo Porsches (Starkey)
Porsche 944 (Long)
Porsche 964, 993 & 996 Data Plate Code Breaker (Streather)
Porsche 993 'King Of Porsche' – The Essential Companion (Streather)
Porsche 996 'Supreme Porsche' – The Essential Companion (Streather)
Porsche 997 2004-2012 'Porsche Excellence' – The Essential Companion (Streather)
Porsche Boxster – The 986 series 1996-2004 (Long)
Porsche Boxster & Cayman – The 987 series (2004-2013) (Long)
Porsche Racing Cars – 1953 to 1975 (Long)
Porsche Racing Cars – 1976 to 2005 (Long)
Porsche - Silver Steeds (Smith)
Porsche – The Rally Story (Meredith)
Porsche: Three Generations of Genius (Meredith)
Powered by Porsche (Smith)
RAC Rally Action! (Gardiner)
Racing Camaros (Holmes)
Racing Colours – Motor Racing Compositions 1908-2009 (Newman)
Racing Mustangs – An International Photographic History 1964-1986 (Holmes)
Rallye Sport Fords: The Inside Story (Moreton)
Roads with a View – England's greatest views and how to find them by road (Corfield)
Runways & Racers (O'Neil)
Sauber-Mercedes – The Group C Racecars 1985-1991 (Starkey)
Schlumpf – The intrigue behind the most beautiful car collection in the world (Op de Weegh & Op de Weegh)
SM – Citroën's Maserati-engined Supercar (Long & Claverol)
So, You want to be a Racing Driver? (Fahy)
Speedway – Auto racing's ghost tracks (Collins & Ireland)
Subaru Impreza: The Road Car And WRC Story (Long)
Supercar, How to build your own (Thompson)
Tatra – The Legacy of Hans Ledwinka, Updated & Enlarged Collector's Edition of 1500 copies (Margolius & Henry)
To Boldly Go – twenty six vehicle designs that dared to be different (Hull)
Toleman Story, The (Hilton)
Toyota Celica & Supra, The Book of Toyota's Sports Coupés (Long)
Toyota MR2 Coupés & Spyders (Long)
Two Summers – The Mercedes-Benz W196R Racing Car (Ackerson)
TWR Story, The – Group A (Hughes & Scott)
TWR's Le Mans Winning Jaguars (Starkey)
Unraced (Collins)
Works MGs, The (Allison & Browning)
Works Minis, The Last (Purves & Brenchley)
Works Rally Mechanic (Moylan)

www.veloce.co.uk

First published in March 2023 by Veloce Publishing Limited, Veloce House, Parkway Farm Business Park, Middle Farm Way, Poundbury, Dorchester DT1 3AR, England.
Tel +44 (0)1305 260068 / Fax 01305 250479 / e-mail info@veloce.co.uk / web www.veloce.co.uk or www.velocebooks.com.
ISBN: 978-1-787117-67-9; UPC: 6-36847-01767-5

© 2023 Johnny Tipler and Veloce Publishing. All rights reserved. With the exception of quoting brief passages for the purpose of review, no part of this publication may be recorded, reproduced or transmitted by any means, including photocopying, without the written permission of Veloce Publishing Ltd. Throughout this book logos, model names and designations, etc, have been used for the purposes of identification, illustration and decoration. Such names are the property of the trademark holder as this is not an official publication. Readers with ideas for automotive books, or books on other transport or related hobby subjects, are invited to write to the editorial director of Veloce Publishing at the above address. British Library Cataloguing in Publication Data – A catalogue record for this book is available from the British Library. Typesetting, design and page make-up all by Veloce Publishing Ltd on Apple Mac. Printed in India by Replika Press.

Lotus Evora at Évora, Portugal. (Courtesy Rui Coelho)

ACKNOWLEDGEMENTS

Firstly, I'm very grateful to Matt Becker for penning the Foreword. As Lotus' Executive Vehicle Dynamics Engineer, Matt was pivotal in establishing the Evora as the paradigm of ride and handling for GT cars, an accolade it's never lost.

Over the years, Lotus' Director of PR, Alastair Florance, has enabled interviews with key individuals involved with the Evora project, from inception to the latest Emira launch, ranging from Roger Becker to Richard Rackham, Matt Becker to Gavan Kershaw, Russell Carr to Steve Crijns, Tony Shute to CEO Mike Kimberley – under whose auspices the Evora came into being – Andy Pleavin and Matt Windle. Alastair also supplied a large quantity of material from the press and publicity archives for me to draw on, and, crucially, a raft of Evora press cars to drive. Lengthy road trips in these cars took me to a variety of locations, ranging from Porto, Verona, Geneva, and Serre Chevalier, to Cheddar Gorge and the Isle of Man.

When seeking advice on the Evora market, independent commentary came from Lotus expert Guy Munday, and I've included Evora-specific interviews with design guru Stephen Bayley, and Goodwood's resident Lotus aficionado, Lord Charles March.

I've also been provided with some stunning images by my valued photographic colleagues and friends, including Antony Fraser, Jason Parnell, Rui Coelho, Alex Denham, Stephanie Ewen, Pedro Domingues, Teresa Cherfas, Laura Hampton, Jules Tipler, Michael Baillie, Max Earey, Kostas Sidiras, Elisa Artioli, Laura Drysdale and Kristie Becker, plus artwork produced specially for this book by Kate Hunt and Sonja Verducci. On more than one occasion, Clive Chapman kindly enabled access to photo locations, as well as classic Lotus F1 cars as photographic accompaniments.

Johnny Tipler

CONTENTS

ACKNOWLEDGEMENTS 4
FOREWORD 6
EVORA TIMELINE 8
INTRODUCTION AND OVERVIEW 10

1 CHALLENGES – Productionising a brand-new car 26
2 STYLE COUNCIL – The rendition of a new model 36
3 SEEKING APPROVAL – From acceptance to sign-off 64
4 UNDERPINNINGS – Keystones of the platform 80
5 DRIVELINE – Powertrain selection and tuning 93
6 DYNAMIC QUEST – Ride and handling 123
7 ASSEMBLY – Going down the line 155
8 ROAD TRIPPING – Driving thrills 179
9 AFTERLIFE – Crowning the successor 223

INDEX 239

FOREWORD
By Matt Becker, Lotus' Executive Vehicle Dynamics Engineer

I started at Lotus in 1988 as an apprentice, straight from school, and once I become a junior engineer in 1994, I spent most of my time working on Elises and Elise variants. So, I knew that platform inside out, and I knew what I liked about all the different Elise-variants we'd done. When Mike Kimberley came on board, we had the opportunity to create the Evora, and in 2007 I was given the chance to lead on the driving dynamics aspects of the car. That was a great opportunity for me, because I had all this knowledge and experience of Elises. Although the platform and the concept of the Elise weren't my projects, I knew a lot about those cars, and what I liked and what I didn't like. The Evora presented the opportunity to take all that knowledge and experience, and work with other key members within the team to turn it into the car it became. That was quite an amazing gift, career-wise.

My dad, Roger, was the Technical Director responsible for engineering – he'd been at Lotus since 1966 – and he and CEO Mike Kimberley got on very well, and Mike trusted dad to achieve the creation of the Evora. In terms of relationships, it was sometimes quite difficult because I was effectively working with my dad. Not that we had a problem with each other, it was just that, occasionally, you had to agree to differ. I worked for him, so it was a work-colleague relationship, but he had forceful opinions because he was very strong in dynamics as well. Sometimes, those conversations became quite difficult because we wouldn't always agree. But the majority of the time, his experience proved him right, whereas I was still learning things.

Matt Becker was largely responsible for honing the Evora's impeccable ride and handling. (Courtesy Jason Parnell)

Any engineer in the car industry would like to take a car from a clean sheet of paper into production, so from concept into production. That doesn't happen very often. I've been fortunate in having that with the Evora and, more recently, the Aston Martin DBX as well; another clean sheet of paper car that I was involved with.

The Evora was actually not a complicated car, being passive with no active or adaptive systems. By having a role in a different company, you quickly realize what complicated cars are: the adaptive systems they have, electric steering, an electronic

FOREWORD

stability programme on another level, and far more complex electronics. Our target from the outset was to get as close as we could to Elise levels of agility, but with much better road isolation and ride performance. Being a driver's car was always the key brief, as per any Lotus, but having greater levels of 'GT' about the car were key, too.

Because the Evora is a passive car, we couldn't rely on any electrical systems to provide purity of control. It was all down to the mechanics, getting the fundamentals of the car right. The correct torsional stiffness, suspension hard-point stiffness, kinematics and compliances were key to this. I've mentioned it not being a complicated car, and that's not negative, because it does delivers purity, as there aren't any electronics trying to interfere, nothing trying to defy physics and electronically adjust the damping in the car. It's absolutely pure.

This is also thanks to the Evora's steering ratio, which is around 16:1, now quite slow in the industry relative to other cars. I spent a lot of time working with TRW to tune the steering system, within which there's basically a torsion bar. You can change the stiffness of this and the valves within the hydraulic system, which adjusts how much effort the steering requires off the centre.

The Evora was targeted at a market sector that would always be challenging for Lotus. If you go up against Porsche in anything, you're going to be challenged – the reliability, quality and usability of its cars are very well-studied, and that's why they continue to sell. So, the Evora was always going to struggle in that marketplace. Also, in 2009, when the car was launched, it happened to coincide exactly with a global recession. The price point was almost head-to-head with some Porsches, like the Cayman, with which it would be competing, especially with the content it had. But we did the best job we could to make the car dynamic. It was as brilliant as it could be, with the budgets and timing we were given.

For me, the Evora project was the pinnacle of my career at Lotus. Not in technical content, as cars have progressed so much since then, but it was a project in which I utilised all of my experience, and, as part of the team, developed something very special. To this day the car is revered as a dynamically brilliant car.

Evora artwork painted by Kate Hunt. (Courtesy Kate Hunt)

EVORA TIMELINE

1999 — Type 118 3.0-litre V6 concept car unveiled, prefiguring the Evora.

2004 — Design work begins on stillborn Esprit replacement, transferred to Project Eagle.

2006 — First product profile signed off.

2007 — Crash testing for Type Approval; 23 prototypes built.

2008 — Type 122: Evora introduced at the NEC on 22nd July, with Toyota 2GR-FE V6 3.5-litre power.
The Evora wins the Royal Automobile Club's Dewar Trophy, based on Lotus' Versatile Vehicle Architecture (VVA) chassis technology.

2009 — Type 122 Evora in production.
CEO Mike Kimberley retires in July due to accident, succeeded by ex-Ferrari Marketing Manager Dany Bahar.

2010 — Evora S is Harrop-supercharged model, but not intercooled.
Type 124: Evora Cup GT4, supercharged 4.0-litre Toyota-Cosworth V6; 30 cars made.
Type 124 Evora GTS and Enduro released in 2011; 6 cars made.

2011 — Evora GTE issued as homologation special for LM GTE and GT4 racing versions; 2 cars made.
One-off Evora 414E Hybrid runs with two electric motors plus Lotus 3-cylinder engine.

2012 — Dany Bahar stands down as CEO.

2013 — Type 122 Evora Sports Racer launched.

2014 — CEO is Jean-Marc Gales.

2015 — Facelifted and Edelbrock-supercharged Evora 400 with intercooler replaces both the standard Evora and Evora S.

2016 — Supercharged Evora GT410 Sport features lighter carbon-fibre panelling.

2017 — Chinese conglomerate Geely acquires a controlling stake in Group Lotus, 24th May. New CEO Feng Qingfeng replaces J-M Gales.
Type 122 Evora GT430 is available: its designation relates to bhp power output.

2018 — Phil Popham is appointed CEO of Lotus Cars, while Feng Qingfeng remains in charge of Group Lotus.

2019 — Type 122 Evora GT is available for US market with 422bhp on tap.

2021 — Matt Windle, previously Engineering Executive Director, becomes Managing Director, Group Lotus.
Type 131 Emira launched, replaces Evora from 2022.

An Evora 400, built 2015, here at Heydon Hall, built 1584. (Courtesy Jason Parnell)

INTRODUCTION AND OVERVIEW

The Evora was Lotus Cars' flagship model from its launch in 2008 until late 2021. It carried forward the mantle of the long-running Esprit, and was itself succeeded by the Emira. Its decade-long production span incorporated only a few mechanical and technical evolutions, and similarly, just a few alterations to the body styling other than enhanced cooling and aero tweaks. Though the launch was smooth, the Evora was soon pitching in choppy waters. The first wave to break was when its patron, CEO Mike Kimberley, had to take early retirement after seriously damaging his back in a fall. The following seven years found Group Lotus in turmoil, pulled this way and that by the spring tides of successive boardroom blitzes, largely propelled by reckless overambition, leading to a drain of engineering talent. Only in 2017 was a sense of calm restored, following the corporate rescue by Zhejiang Geely Holdings, which owns 51 per cent of Lotus Cars.

The Evora rode out these turbulent seas, manifesting a couple of aerodynamically driven face-lifts, maxing out with the GT410 Sport model of 2018. That's the broad background; we'll get into the conception, development construction and specifications in subsequent chapters, culminating in on-road experiences with various Evora cars, plus a look at the Evora's successor, the Emira.

Characterising the Evora is not straightforward. To get back to basics for a moment, there are two-seater sportscars such as the Elise family, and there are four-seater varieties usually described as 2+2s: Grand Touring cars like the Aston Martin DB9 Vantage, Jaguar F-Type, Maserati GranTurismo, Ferrari Roma, and Porsche 911. In this sense, the two-seater Exige coupé is comparable with the two-seater Porsche Cayman, Alfa Romeo 8C and Alpine A110. The Lotus Evora is essentially a 2+2, but it's more sports car than grand tourer, a blend of avant-garde styling and ingenious engineering by Lotus Cars, powered throughout its lifespan by the Toyota V6 engine. It was based on a new generation of extruded and bonded aluminium chassis, placing it in a long tradition of ground-breaking Lotuses, from the

All outings in Evoras are thrilling, and mine to Verona certainly had its moments, as revealed later in the book. (Courtesy Jason Panell)

fibreglass '50s monocoque Elite, the Seven that brought race car dynamics to the road, to the '60s style-icon Elan and the Esprit supercar, which showed Hethel could match Maranello, Bologna and Stuttgart.

Leading up to the Evora's arrival in the Lotus line-up, we should revert to the inception of the Elise, which ushers in 'the modern era.' Step forward Romano Artioli, whose Bugatti concern bought Lotus Cars from General Motors in 1993. Snr Artioli relinquished control in 1996, with ownership passing to Malaysian car maker Proton in early 1997, though Artioli remained at Lotus as Special Projects Director till 1998.

Enter Mike Kimberley. Having spent the best part of 22 years at Lotus, working directly for and with Colin Chapman from the late 1960s, Mike Kimberley rejoined the supervisory board of Group Lotus in 2005, and was reappointed CEO in May 2006. Brutally, he was obliged to retire early in 2009, due to breaking his back in a tumble in an icy car park – although not before he'd overseen the conception and launch of the Evora. He was also responsible for re-establishing Lotus Engineering as a global high-tech business, and restored the Lotus Group to financial stability.

From August 2006 to 2008, the Hethel plant was engrossed with the design, development and production of the brand-new Type 122 Evora. Conceived by Mike Kimberley as the first step in his five-year business plan, the Evora was unveiled at London's Excel Arena in 2008. The Evora was a blend of avant-garde styling masterminded by Head of Design Russell Carr and ingenious engineering hatched by Richard Rackham. The Evora chassis shared the bonded aluminium technology pioneered on the Elise, though comprised three independent modules, including a central tub and add-on sections front and rear. Some body elements were also stressed members, which boosted stiffness, and the underside was flat-bottomed for optimum aerodynamic airflow.

Aerodynamic efficiency establishes a low drag coefficient, avoiding the need for a large engine that normally brings a weight penalty, and thus costs more to achieve the desired speed and handling goals. That had always been a Lotus principal, from the earliest days, born of Colin Chapman's training in the aircraft industry. The science of the low-drag design involves smooth body sides to keep the air attached – as opposed to a high drag 'Coke-bottle' body side that shrugs off the airflow. The teardrop-shaped cabin and glazed rear window over the engine bay also help keep the air attached, contrary to the tunnel-back configuration with flying buttresses that interrupts the airflow. In the original specification, the powerful bi-xenon headlamp clusters were specially made, and upgrade packs were available that featured parking sensors and the much-vaunted rear-view camera. Folding mirrors were installed because the car is relatively wide, and in some markets, such as Japan, electrically-operated mirrors are an important consideration when parking and leaving the car.

The Evora's suspension incorporates components fabricated in forged aluminium, fixed to a chassis structure composed of bonded aluminium extrusions and a subframe made by Lotus Lightweight Structures in Worcester. Formerly Denmark-based Hydro Aluminium, where the first Elise chassis were made, the plant was acquired by Lotus in 2008, and in 2020 the facility relocated to Hurricane Way in Norwich's light industrial hinterland, close to the city's airport. The sports suspension componentry includes wishbones that are as light as the Elise's, but twice as stiff, and were very expensive to develop. The Evora is longer than the Elise, but the intention was always that it should provide the same standard of Lotus handling.

The Elise is familiar enough to provide intelligible contrasts with the Evora, so while both the wheelbase and track on the Elise are significantly shorter than the Evora's, and although the chassis looks large, the car's width has been controlled by cleverly reducing the sill width, yet retaining its strength. That crucial contact point with the road surface, the tyre, was the subject of intense development in conjunction with Lotus' regular suppliers, Yokohama, though eventually an off-the-shelf Pirelli P-Zero tyre was selected for Evora. Meanwhile, Lotus engineers perfected a Bosch ABS system that matches the car's performance.

An Evora media outing visits the Fiat plant's rooftop test track at Lingotto, Turin, 2011. (Courtesy Antony Fraser)

The Elise's ride, handling and roadholding is so good that a standard car could run in a road-race like La Carrera Panamericana, demonstrated here by Rachel Larratt and Steve Warwick deep in the Mexican jungle in 2006. (Courtesy Johnny Tipler)

The Evora's 3.5-litre Toyota V6 powerplant was monitored by Lotus' own engine management system to produce 276bhp, in original normally-aspirated format, driving through a six-speed Toyota gearbox. Despite a relatively modest power output, a sub-1400kg kerb weight helped achieve a decent top speed of 162mph and a 0-60mph time of 4.9 seconds. Then, in 2010, the Evora S was equipped with a Harrop supercharger, but no intercooler, and another major evolution in Evora spec came in 2015 with the face-lifted and Edelbrock-supercharged Evora

From top: Evora at Le Domaine du Mas de Pierre, St-Paul-de-Vence, 2010, for Provençale media session. (Courtesy Antony Fraser)

Pausing for a cuppa at Heydon village tearoom, the Evora 400 of 2015 featured aerodynamic bodywork revisions. (Courtesy Jason Parnell)

The Evora gained a supercharger in 2010, identified by the S suffix. (Courtesy Jason Parnell)

Neat, rational and classy: cockpit of the 2015 Evora 400. (Courtesy Jason Parnell)

400, now with intercooler, replacing both the standard Evora and Evora S.

At that point, the Evora cabin was the most comfortable interior that Lotus had ever created and, obviously, it's far easier to get in and out of than an Elise, though the Type 121 Europa, with its cutaway sills and taller roofline, offers a compromise here in the two-seater accessibility stakes. Because the Evora was not originally intended to be a trackday car – though it was only a couple of years before a race version was offered – it can afford a higher opening line at the top of the door so you don't need to duck to get in. Similarly, the sill is lower and narrower and the seat sits some 65mm higher off the floor than in the Elise.

The interior echoes the lines of the exterior, having a floating centre console and a contrasting leather band running as a hoop from the dash right around the doors, cabin sides and across the rear seats. The Recaro seats are way more substantial than those in the Elise, and, due to cabin size, there's more shoulder room between the driver and front seat passenger. A lever hidden away in one of the portals at shoulder level in the seatback hinges the seat forward to allow access to the rear of the cabin. The dash is covered in leather and aluminium, switches are edge-lit and an optional central touch-screen can operate radio, iPod/mp3 and sat-nav controls.

The Evora took the then-current Elise two-seater family onto another level, providing an opulent cabin with two occasional rear seats, yet retaining the mid-engined layout for optimum balance. Subsequently, in 2013, the Exige wheelbase was lengthened and the Toyota V6 engine installed, placing it more on a par with (but dynamically quite different to) the Evora.

Going green: Laura Hampton snaps the Exige 430Sport at Bintree Mill, Norfolk. (Courtesy Johnny Tipler)

The Evora's 2+2 cabin is spacious enough to accommodate two adults and two children – or a load of luggage. (Courtesy Lotus Cars)

Although it was possible to specify no seats in the Evora's rear cabin – sometimes known as the 2+0 layout – the 2+2 cabin provides space for a pair of adults in the front and a couple of children in the back. Or, as I found, a youth (my son Alfie) stretched from side-to-side. Whether you have the seats or not, the space it provides means a good size supermarket shop, a round of golf, or even a drum kit (!) is more than just a faint possibility.

Significantly, the Evora had the blessing of the Chapman family, especially Clive and his mother Hazel, and that's a crucial benefit to any Lotus model because their endorsement represents continuity with the marque's heritage. Speaking at the Lotus Owners Group convention (LOG28) in Indianapolis in mid-2008, Clive Chapman commented, "It really is genuinely a Lotus: you only have to look at the neatness of the package to see that. A journalist asked me what my father would have thought about the car, and I said that if he saw that chassis he really would be excited. The shape of the car is fantastic but, take the body off, and what's underneath truly is a Lotus."

By any standards, the Evora's gestation was extremely rapid. The initial renderings were sketched in August 2006, and the first prototypes were being tested in early 2008. Visiting the Bosch test centre at a snowy Arjeplog in Swedish Lapland in April that year, I rode on the frozen lake proving ground in a camouflaged 'Eagle' chassis that was presented as an extended Esprit mule. The disguises came off the prototypes after the unveiling in July 2008, and VP (verification prototypes) were being constructed in October. The parts-in-plant date was September 24th, with the first cars scheduled for build in December 2008. The first 16 production cars were earmarked as prototypes on which to develop the US Federal version, with a view to productionising that in October 2009.

Although development progressed rapidly, the Evora's incubation was not at the expense of the other Lotus models. With 100 engineers and technicians flat-out on the Evora, there were still 50 engineers working on the small-platform Elise, Exige, Europa and 2-Eleven. Vehicle Engineering Director Roger Becker was another driving force.

He had been at Lotus for 42 years, and seen many projects come and go. "We'd had a bit of a dry spell for the past 13 years," he commented during the Evora launch period. "The last totally brand-new car we did was the Elise. That burst onto the scene and surprised everybody, because it was a product that took Lotus back to its grass roots." The big cars – the Esprit V8s – had probably had their day, and it was time to revisit smaller,

Clive Chapman chats to Roger Becker in the 2-Eleven outside the Indianapolis Hilton during LOG28 at Indianapolis, 2008. (Courtesy Johnny Tipler)

At close of play in 2021, the Lotus line-up comprised Exige 430Cup, Evora GT and Elise Sport240. (Courtesy Lotus Cars)

nimble cars like the Elan, which was under development when Roger Becker joined the company in 1966. "They were fantastic cars," he said. "And I should know, because I broke a fair few in testing. But now, [in 2008] the time is ripe for us to launch another bigger car – the 2+2 Evora."

The name game
Lotus' best-known model names have stood the test of time well enough to merit recycling – the Elite, Elan, Europa – and eventually there'll probably be a new Esprit. But the Evora received a brand-new name because, like Elise, the product was sufficiently different to warrant a distinctive moniker all of own.

Lotus' project cars are generally codenamed after race circuits during the pre-production phase, as in 'Seca' for the Elise S, 'Atlanta' for the Exige S, 'Croft' for the Elise R, 'Imola' for the Elise SC, and 'Sepang' for the Europa, and by no means do all come to fruition. But an exception was 'Eagle' for the Evora. True, not a racetrack, but although many anticipated the Eagle name would stick, it was never actually in the running as the definitive choice. The Eagle brand-name that ran as the All-American Racers' F1 car in the late 1960s is owned by the estate of Dan Gurney, who may have been Lotus hero Jim Clark's sparring partner in the '60s, but was still fiercely protective of his own car's name. Indeed, American Eagle kit cars and Eagle motorcycles and other two-wheel interests using derivatives of the Eagle word – were also potentially litigious. In the early 1990s, Eagle was a marque of Chrysler Corporation following its purchase of American Motors Corporation (AMC). Its Eagle was aimed at the enthusiast and although the brand was short-lived, they sold more than 115,000 units of the Talon coupé. Clearly, though, the Eagle name already carried too much baggage in the automotive world: another word was needed.

Not only was Evora an E-word, it was an inspired choice, redolent of the exotic. Évora is an ancient city, a World Heritage site about an hour-and-a-half's drive east of Lisbon in Portugal's central Alentejo region. With a 12th century cathedral and 1st century Roman temple at its medieval heart, and a bustling university its modern powerhouse, Évora is an apt combination of tradition and innovation. Apart from the Portuguese connections – the eponymous City, the Cabo Verde folk singer and the Olympic long jumper – the word Evora even contains elements of other inspirational words: evolution, evoke and aura. Product Planning Manager Rob Savin explained the model naming strategy. "Traditionally, Lotus model names begin with 'E.' We had a long list of 'E' names that had been proposed before, and we looked at those again. We also put together an algorithm that created lots of fictitious 'E' words, and we looked through dictionaries and atlases for obscure ones. The name has to work in all countries: it mustn't sound silly or be a swear word in another language, and it needs to be available in the sense that we can obtain trademarks for it." Lotus took a shortlist of 20 possible handles, and tried them out in different

The Evora 410Sport visits its namesake: the Portuguese city of Évora, a World Heritage site 130km east of Lisbon. (Courtesy Rui Coelho)

languages. The core naming team, consisting of Roger Becker, Tony Shute, Rob Savin, Matthew Jones, Russell Carr and Katie Dann, who was Senior Executive, Licensing and Trade Marks, pruned the shortlist to ten. "We proposed our final choices to Mike Kimberley and the Proton chairman for their approval, and they went for our favourite name, the Evora," said Rob.

According to Vehicle Engineering Director Roger Becker, "while the Elise was built like a Meccano set, bolting bits on as it went down the production line," the Evora was composed of a number of modules. The front module was assembled off the line, incorporating the suspension, brakes, the cooling system and steering assembly. While the central chassis tub was going down the line having the wiring and mechanical components added to it, the front module was being created a short distance away on a sub-assembly line.

Assembling the wiring for the Evora dash cluster. (Courtesy Lotus Cars)

On the Evora assembly line, the Toyota V6 engine and gearbox are installed in the chassis' rear module. (Courtesy Lotus Cars)

Likewise, the rear module comprised the subframe complete with engine, suspension, brakes and exhaust. The seatbelt anchorage frame was built up as a separate unit, and when the tub went down the line there came a point where the front and back modules were bolted onto it, and then the body components were attached as well. It used a conventional hydraulic power-steering system, necessary because the of Evora's relatively broad tyre width – plus the target market demanded it; an electric power-steering system was considered but it was felt to not offer enough 'feel' to the driver. The seats were top-of-the-range Recaros. The 2+2 theme is often defined by the GT suffix, since the role of the 2+2 cabin is to accommodate children, or smallish adults on a road trip, in some style, with additional carrying capacity. Admittedly, the Evora, like its smaller siblings, is not so well endowed in that respect, but when we get into the drive stories I will testify that lengthy journeys can be undertaken with adequate luggage for two.

The concept of the grand touring car originated in the lavishly-styled Grand Routiére extravaganzas that graced the smart Riviera resorts in the 1930s, and the breed has permeated the entire spectrum of sports car motoring, from gentleman's carriages of the '50s and '60s, like Bristol, Jensen and Lancia, to Aston Martin, Bentley and Jaguar, and big boys' toys from Mercedes-Benz, Porsche and BMW. Thanks to the fine tuning its suspension received in the development phase, the Lotus Evora is rather more of a nimble sports car than grand tourer.

Archetype of the Grande Routière segment: the 3.5-litre straight-six 1937 Delahaye 135M, with coachwork by Figoni et Falaschi. (Courtesy Jason Parnell)

As we shall see when putting it into practice, it's also possible to describe it as a grown-up Elise, partly due to its mid-engined layout. Yet it can still seat four people – at a push.

For the drivers
To borrow the company's most recent adage, 'For the drivers,' ideologically, Lotus products have always been about the driving experience, and never just about getting from A to B. The

At Classic Team Lotus' Hethel premises, the Type 72 F1 car and Evora GT410 Sport embody the marque's historical production of road and racing cars. (Courtesy Stephanie Ewen)

Designed by Marcello Gandini for Bertone, the Ferrari Dino 308 GT4 (1973-1980) located its 3.0 V8 amidships. (Courtesy Johnny Tipler)

company's origins on the racetrack, with Team Lotus' seven F1 World Constructor's Championships and innumerable race and class victories achieved by work's drivers and privateers over a 70-year history, are manifest in the Evora.

A fusion of lessons learned and concepts honed in previous sport and racing models, it's a paragon of 'form following function.' Its elegant, purposeful aerodynamic lines clad its state-of-the-art chassis, running gear and powertrain. Easily removable structures facilitate repair and replacement, while accessibility for routine servicing is reasonably practical. The majority of its components are sourced within Lotus' existing suppliers, which is an advantage in quality control as well as speed of development.

Project Eagle chief Tony Shute remarked, "2+2s are not easy to do." A prime mover behind the company's mid-1990s Elise-led renaissance, Tony had already spent 22 years at Lotus when tasked with developing the Evora. "With a mid-mounted engine, it's quite a sophisticated conjuring job to get four seats into a coupé shell, plus the space behind them to stow a set of golf clubs or whatever." A handful of makers, Lotus included, have built 2+2 sportscars in the past, but just four have placed the engine amidships and still squeezed a pair of seats in the back. As most of us know, the flat-six engine of the Porsche 911 lives behind the rear axle, facilitating a 2+2 cabin.

Probably the best-known example of a mid-engined sports car is Ferrari's Dino 308 GT4, from Bertone's design studio and in production from 1973 to 1980. Its contemporary, the Lamborghini Uracco from 1972-79, was penned by Marcello Gandini and was a similar take on the Dino. The V6-engined Maserati Merak of 1972-82 by Giorgetto Giugiaro was simply a reworked version of uts V8 Bora, with two rear seats squeezed into the space created by the use of the shorter engine, and the Ferrari Mondial built from 1980 to 1992 demonstrated that even Pininfarina had

Being a 2+2, for ten years, this 3.6-litre Porsche 964 served on Tipler's daily 70-mile school run as well as numerous family journeys to the Douro in northern Portugal. (Courtesy Johnny Tipler)

a hard job harmonising sports GT lines with a 2+2 cabin. And that's it – until the Lotus Evora.

Tony Shute reflected that, "Aesthetically, the most successful example of the classic mid-engined 2+2s was the Ferrari 308 GT4, but we could hardly justify doing it on aesthetics alone. It took a lot of debate, because it's really a question of whether you can rationalize making a 2+2. Is there a need for it, and is there a market? In my view, the reason why the Porsche 911 has stood the test of time – beside the fact that it's a very good car – is that you've got space in the back to take your mate down the pub or whatever, just for a short journey, and you can put a wet rain-coat or your shopping on the back seat – or even a couple of small children." As I did for some years in 911s, on the 30-miles-each-way school run from Norwich to Leiston, with Zoë and Alfie, plus the dog, twice a day.

Product Planning Manager Rob Savin, who worked with Tony Shute, evaluated competitors and Lotus product target assessments. Part of Rob's role was to represent sales and marketing, PR, and the commercial aspect of the new model. He wrote the product profile and defined the car, identifying its market and estimating sales figures. He also commissioned the external market analysis agency whose research confirmed his findings. As the car launched, his task was to ensure that the business case proved to be correct. He cited the Audi TT as another good example of a front-engined 2+2, though getting into it was much more troublesome than the Evora. "It's a sophisticated conjuring trick to get four seats in a coupé shell, plus the stowage space behind them," said Tony. "You mustn't compromise the car by giving it too much accommodation in the back seats. The key thing is to provide a degree of practicality. In the UK, the average number of occupants per car is 1.4, and most of the time cars are occupied by a single person, so for people who say, 'I can't get four big people in there,' this is not the car for them. There are some perfectly good four- or five-seater performance cars out there, which is what you need for everyday transport if you've got two kids and a dog. But for a car that you can live with, day to day, that area behind the seats totally transforms the proposition." And, as I've said, I can totally identify with that. I've even given up motorbikes because of it ...

The Evora enjoyed a distinctive role within the Lotus line-up. For any car designer, its proportions are essential to a successful design. The delicate balance between length, height and width, the size of front and rear overhangs, the cabin position and the diameter of wheels can determine the stance of the vehicle, and whether it looks lithe and agile or heavy and ponderous. The execution of line, surface and detail are important, but without the basic framework that the proportions provide, it is doomed to failure. This is even more true with a sports car, and therefore the additional challenge of accommodating a plus-two capability was particularly challenging for the Evora design team.

LOTUS EVORA

Amongst the core naming team were Morris Dowton, Head of Manufacturing for the original Elise, and Katie Dann, Senior Executive for Licensing and Trademarks (Courtesy Richard Parramint)

The Evora's coupé shape looks gorgeous from any angle. (Courtesy Antony Fraser)

Emotional rescue

Most people are influenced when they buy a sports car by irrational emotions provoked by the way the car looks. In that case, the car must be aesthetically pleasing in order to seduce the customer, but this can't be achieved at the expense of occupant space, technical hardware or legal criteria. It's a given that a Lotus should look low, sleek and dynamic, even when standing still. In the case of the Evora, the key to accommodating these potentially conflicting requirements within a technically relevant solution was a close working relationship between the studio and the packaging team, led by the Vehicle Architect, Richard Rackham.

While Richard and key colleagues like Andy Pleavin and Russell Carr are still in post at Lotus, many of the main players involved with the Evora during the past 25 years, and whose voices we hear in this book, are no longer working for the company. Why? There's natural career progression, of course, and employees retire. On the whole, though, people were local and tended to stick at Lotus for the duration – former Team Lotus chief mechanic Bob Dance joined (at Cheshunt) in 1960, and still works for Clive Chapman over the road at Classic Team Lotus. But in the case of Evora staff moving on, it was mostly because they were either ousted, or chose to leave, during the anarchic period between 2009 and 2012, when Dany Bahar was CEO, and until 2017 during the lacklustre period when his replacement Jean-Marc Gales was in charge.

That sweeping statement needs clarifying: prior to heading Lotus, Mr Bahar had successfully launched the Red Bull and Toro Rosso teams in Formula 1 and had been Senior Vice President for Commercial and Brand at Ferrari, and was recruited by Lotus Cars' parent company, Proton, when Mike Kimberley was forced into retirement. While Mr Bahar's radical personal ambitions centred on the launch of five brand-new Lotus models displayed at the 2012 Geneva Show, and what *Autocar* described at the time as "poor financial management," the highly lucrative Lotus Engineering consultancy, a Chapman legacy fostered by Mike Kimberley, was squandered and its talented technicians forsaken. Someone told me Dany was "Very good at micro, but less good about macro," implying that he excelled at the fine detail but his big ideas were less successful. Someone else who'd worked under both Mike Kimberley and Dany Bahar said that it had been an honour to work for Mike Kimberley, but that Dany Bahar and his cohort did not like the Evora and had simply "ignored everyone."

From 2007 to 2014, Matt Becker was at the forefront of Evora development, and he provided his personal take on the Bahar years. "In 2010, Dany Bahar turned up. Dany had all these

Former Team Lotus chief mechanic Bob Dance joined the firm in 1960, and still works for Classic Team Lotus, as here at Spa-Francorchamps in 2017. (Courtesy Johnny Tipler)

grand ideas about all these new cars he wanted us to do. And almost all of the existing cars, he put them to one side. So, the Elise, Exige, Evora, he wasn't interested. And then, I think it suddenly dawned on him that you have to be interested in those cars because, otherwise, how are you going to pay the bills between the point you are today and the future? You need to pay attention to the existing cars, because otherwise you can't pay the bills. And it took him quite a long time to realize that. Suddenly, there was all this attention going on the Evora, to actually make it up to quality, make the gearshift better, make it faster. And he started throwing the kitchen sink at the car. But by then, it was almost too late, because he'd put all of his focus and attention – and media attention and company attention – onto these seven new cars we were going to do in seven years.

"When he turned up, at the time, we were like, 'Wow. This is unbelievable. This is all really going to happen.' But people like my father said, 'This is never going to happen.' In 2010, when I think he was 64, dad retired. He said, 'I'm not up for this, because I don't believe it can happen; something this size, with the budgets we have.' So, basically, he bailed out. Dany Bahar gave a presentation to the whole company, basically saying, 'This is my vision. This is what I want to do. If you're on my bus, you're on my bus. If you don't want to be on the bus, then I suggest you go and see Kerry Dugan in HR, and basically find a method of exiting the company.' And that was his motivational speech, which we all felt was pretty poor. That wouldn't have impressed local folks on the payroll. And then, two years later, the whole thing fell apart: the parties that were happening, the gifts that people were getting – including free cars, the helicopters coming in and out of the turning circle, which is opposite the office where we were based. All different sorts of 'stars' coming off the helicopters. And you knew it wouldn't last." During the Bahar interregnum, a number of key individuals were sidelined as his favoured acolytes usurped them. "There was a two-dimensional culture," recalled Matt: "a blend of Italians and Germans, which was about 60/40." Matt kept his counsel, but sympathised with deposed colleagues. "I admire their resilience in staying on at the company, for whatever reasons. But yeah, for sure, it would have been tough as an individual to accept that, when you've been head of a department, suddenly somebody else comes in, and they're getting paid four or five times a year more than you. It's the most demoralizing thing ever. And I guess you either put up and shut up and stick with it, or you do what we've done now, you move around the industry a bit."

When Mr Bahar left, the Malaysian government, which owned Proton, transferred control of Lotus to another of its state-owned conglomerates, DRB-HiCom, who knew rather less about the sports car market, and imposed a year or so's due-diligence on Lotus to ascertain what had taken place during the previous three years. Very few cars were made during this bleak period, and component suppliers were somewhat compromised. Thereafter, Mr Gales was engaged to straighten the ship, so to speak, and belts were further tightened in a quest for profitability, some say, to the further detriment to morale.

There was a knock-on effect on the Evora, too. You might think the 400 model that launched in 2015 was a subtle upgrade to looks as well as performance, but Matt isn't convinced. "It was another strange time in Lotus's history. And that's why the Evora never really got the chance or opportunity to succeed to the level of what it could've been, because you had the global recession. Then you had Dany Bahar turning up and the car was forgotten about, and then it was like, 'Okay, we'd better actually pay some interest to this.' And then, at the point he left, it was then, I think, that the car took a slightly different direction from what it was originally intended to be. It was also face-lifted in the interests of cooling and aero, which made it more aggressive-looking. I was working on the 400 before I left, so we were developing that. And that was part of the changes we made to the car to make it dynamically better. Then, Jean-Marc Gales got involved, and he was massively interested in lap times (around the test track), just obsessed with lap times. I don't think the Evora road cars have ever really been about that. And lap time is obviously important for racing. Perhaps it's important for trackdays. But Lotus would never compromise the feel of the car for a lap time. And I think that's where the Evora started to take on some of these portents. The car just started to get quite aggressive, not just in appearance, but in terms of tweaking the handling to make it achieve those sort of lap times. The original concept was for it to be a more comfortable GT car, to retain the agility of the Elise. Obviously, incorporating high levels of comfort and noise isolation to make it more of a usable car. But you didn't need to make an Evora overly aggressive, because you had the Exige for that."

In 2017, the Chinese conglomerate Geely acquired a controlling stake in Group Lotus, and CEO Feng Qingfeng replaced J-M Gales. I can't help thinking that anyone who hung on till Geely saved the day is something of a superhero or, as someone remarked, knew where the bodies were buried. Meanwhile, to name but a few, people like Tony Shute, Rob Savin, Steve Crijns, Matt Becker – and, sadly, Roger Becker who died – plus a number of other contributors to the Evora project are long out of the picture, though opportunely replaced in the intensive talent recruitment drive rolled out under the Geely regime.

Going back to 2007, then, a packaging revolution was required to deliver the Evora, distanced from the Elise line up, and the team on the Evora programme clearly understood the need to give the studio the freedom it needed, and therefore worked tirelessly to package every component in the most space efficient manner. This not only gave the studio room to express the design, but also helped reduce weight by minimising the overall dimensions. Understandably, the Evora is larger and more imposing than the Elise family, which was seen as being to its advantage in marketing terms. As Tony Shute commented at the time, "in some markets, like the USA, the Elise looks quite small, and yet it is relatively expensive compared with a lot of American offerings, so purchasers have to really understand what its merits are, whereas the more lifestyle customers expect a slightly bigger offering in that price range. In the emerging

An early press car lent to JT when he lived in Norwich Cathedral Close. (Courtesy Johnny Tipler)

markets, in places like China, it's even more difficult, because the more you pay, the bigger it should be, and the more doors you should have."

Size matters, then; another maker facing the same 'more metal for the money' dilemma is Porsche, whose 911 sales in China are negligible, maybe 200 cars a year, whereas its Cayenne SUV sells in thousands. "Bentley sells more cars in China than they do in the UK," Tony claimed; "it's the same in Korea and Malaysia; lots of people want a really impressive-looking car, which the Evora certainly is."

Focused on driver enjoyment, the Elise was the smallest car in its class, but was a traditional sports car incorporating modern technology and performance. The Evora was the car for enthusiasts who were keen on the image and fun of a Lotus, but wanted a more mainstream car in terms of size, interior luxury and electronics. From Tony Shute's point of view, it was a continual balancing act between comfort and weight: "When you add luxuries, things like satellite navigation, decent air-conditioning and hi-fi you are adding weight and getting away from the purity of the driving experience. Today, everyone expects satnav; they've given up looking at paper maps. They want a Bluetooth connection to their mobile phone. The Evora's got all the ancillaries for a modern lifestyle, but it doesn't need seats that weigh 100 kilos because they've got 68 motors in them

The roads criss-crossing Salisbury Plain provide an exhilarating playground for the Evora. (Courtesy Antony Fraser)

controlling every nuance of posture. Instead you get a really good seat that fits nicely and adjusts forwards and backwards and integrates with great cabin ergonomics."

According to Matt Becker, another reason for creating the Evora as a 2+2 was socioeconomic. "It needed to have the rear occupancy because it was always thought that the husband who wanted to buy the Evora needed the extra ability to say to his wife or his girlfriend: 'Yeah, but it has got two seats in the back.' The fact that he couldn't use them was where that fell a bit short. But it was a balance between handling, in terms of wheelbase, and to give it rear occupancy. Storage hardly entered into the picture. And it is noticeable when you drive even a Boxster on a day-to-day basis, you've got storage in the back and storage in the front – at least twice as much storage as in any Lotus, really." Matt was candid: "There's no reason why an Evora couldn't have had a baggage space up front as well, because there's a potential space above where the radiator sits. And they could have put a radiator on each side, like Porsche, and created a trunk space in front."

Here, then, we discover how all those concepts came to fruition, and how the product came to life on the open road.

Johnny Tipler

CHAPTER 1

CHALLENGES
Productionising a brand-new car

A sign of things to come: the 3.0 V6 Type 118 M250 concept car in the Design Studio in 1999. (Courtesy Lotus Cars)

The Evora's gestation started in 2004, when Lotus Engineering began developing a replacement for the long-lived Esprit. A glimpse of what might be around the corner had already been revealed in 1999, in the shape of the gullwing-doored Type 118, AKA the M250 show car, said to be fitted with an unbranded 3.0-litre V6 and transverse six-speed gearbox. A similar question-mark hung over the Evora, initially, when a suitable engine wasn't available. But then, as project head Tony Shute explained, "We found we had an engine that would do nicely for a mid-range car – the 3.5-litre Toyota V6 – so we could justify going from an Elise via the stepping-stone middle segment, and then into the top-of-the-range." Product Planning Manager Rob Savin commented, "We'd already done a lot of work on what was going to be the new Esprit, and we'd learned quite a lot from that, and the opportunity came up to do this mid-range car. It certainly made sense from a marketing standpoint, so we were able to draw some of the lessons out of working on the stillborn Esprit and apply them straight away. People from the design team came straight from that programme as well. The extruded aluminium chassis is one of our core competences, and we were able to produce a new version of that concept by applying what we'd learned in the 13 years since we pioneered the technology with the Elise."

At the core of the Evora's development was the strong

CHALLENGES

relationship with Toyota, whose reputation for quality and bulletproof engineering was crucial to the success of the project, since it provided the powertrain. The choice of the Toyota 3.5-litre V6 as the Evora's engine was a pragmatic one. The Esprit V8 engine would have needed a lot of work to upgrade its emission standards, and was expensive to manufacture. As Platform Manager Tony Shute said: "It's unfortunate that the cost of powertrain systems is very high, so although we can design and develop them for other people, ironically, it's too expensive to use them in our own products. You need really big volumes and timescales to amortise the costs and, with Lotus, the volume is never going to be huge, so to have your own bespoke engine makes the project incredibly expensive."

The Toyota V6 engine powering the Evora is mated to a Lotus-engineered gearbox, clutch and flywheel, plus the ECU, inlet and exhaust, all of which conspire to change the characteristics of the engine, amply justifying the Lotus logo on the cam covers. The Toyota engine has always provided good performance, and is relatively easy and cheap to service. Iconic makes like Morgan, Bristol, Jensen, De Tomaso and Iso have all employed bought-in German or American power-plants and, for a certain amount of its history, Lotus relied on a succession of Coventry-Climax, Ford, General Motors, Isuzu and, more recently, Toyota engines. As Rob Savin pointed out, "Many of the engines in our production cars were adapted or converted units from mainstream manufacturers, going back to the Climax and BMC A-series engines in the Sevens and Elites. We'd tweak them and give them the characteristics that suited our cars, and that holds true today with the supercharged Elise and Exige S, and the Europa S with its turbocharged GM engine. We've really made that Toyota engine our own."

The leader

Managing Director and CEO Mike Kimberley joined Lotus in 1969, tasked with improving the existing model range and running continuous engineering. His career at Lotus almost finished before it had begun, when he took a Bond Bug three-wheeler home, which had been gifted to Colin Chapman, and wrote it off on the way. The Guv'nor fired him – and reinstated him two days later! Mike's first project in 1970 was the Europa Twin-cam, engineered, developed and productionised in a mere 13 months. Delighted, Colin Chapman awarded him 500 shares as a reward. Mike then became Vehicle Engineering Manager in 1972, Chief Engineer in 1974, and Engineering Director in 1975. In March 1976, Mike became Operations Director, and was appointed Managing Director of Lotus Cars in October 1976. The new Lotus models during Chapman's move upmarket were the

The long-term rapport and synergy with Toyota facilitated adoption of the 3.5-litre V6 engine, subsequently supercharged as in the 400 model shown here at Heydon, Norfolk. (Courtesy Jason Parnell)

Elite M50, the Eclat M55, and the Esprit M70. Kimberley and Chapman were a well-matched dynamic duo, a blend of Mike's astute business acumen and disciplined engineering approach, and Colin's maverick innovative design and engineering vision. Colin and Mike initiated Lotus Engineering in January 1977 as a consultancy similar to that of Porsche at Weissach. Lotus Engineering provided third-party clients with advanced Lotus-developed technologies. These included Lotus' own all-aluminium four-valves-per-cylinder engine, high-performance lightweight advanced combustion and performance technologies, unique VARI composite structural body and component closed-matched tool body systems, plus global benchmark vehicle dynamics, active controls and anti-noise, not forgetting a continuous cascade of advanced technology deriving from Team Lotus in Formula 1.

A little bit more background on Mike Kimberley is in order, presaging the Evora, because it helps get us up to speed on the corporate history. Over the years, Lotus engineered and developed specific models for a raft of automotive firms, including Chrysler, Ford, General Motors, Isuzu, Vauxhall, Aston-Martin, Kia and Hyundai, as well as governments and military, ranging from the Antarctic tracked Snow-Cat to tank suspensions for the military, and ultra-lightweight competition bicycles.

Colin Chapman and Mike Kimberley developed clients independently – Chapman through Group Lotus and MJK through Lotus Cars. Mike began the Talbot Sunbeam Lotus project in January 1977, creating the basis for a successful rally car, using the Lotus-developed 2.2-litre dry-sump twin-cam engine, leading to the production of road cars for sale. In 1981 the Talbot Lotus Sunbeam won the Manufacturers' title in the World Rally Championship and 1982 RAC Rally. This was followed by the Group Lotus initiated DeLorean DMC-12, and, in 1990, Mike, Jack Smith and Bob Eaton pushed through the GM Lotus Omega/Lotus Carlton programme against a very reluctant Opel.

In 1980, Mike set up a close co-operation and technology agreement between Lotus and Toyota to provide the 'big brother'

Godfather of the Evora, Group Lotus CEO Mike Kimberley pictured in November 2008. (Courtesy Lotus Cars)

Three key figures in the Evora gestation: Roger Becker, Luke Bennett, Mike Kimberley, 11th September 2008. (Courtesy Lotus Cars)

that Lotus needed to re-enter the Elan higher volume/lower priced market. Colin and he signed the agreement in Japan in 1980. After Colin Chapman died of a heart attack in December 1982, Mike was appointed Group Chief Executive, working very closely with Toyota to enable Lotus to survive the crucial next six months, and ensuring a small recapitalisation and stabilisation of the company in July 1983, which, controversially, engendered negative results for the Lotus car business when considered over the medium to long-term.

In January 1986, General Motors, following abnormal secret majority shareholder actions, took a controlling interest in Group Lotus, buying out Toyota's share in the process. GM was a very good stakeholder, investing heavily in modernising Lotus Cars' plant, equipment and facilities. Mike acquired Millbrook Proving Ground for Lotus Engineering in 1987, whilst a member of various GM strategy boards and executive committee meetings. In January 1992, Mike was appointed Executive Vice President of GM Overseas Corporation and relocated to Kuala Lumpur, Hong Kong and then Singapore. Next, he was headhunted for President and CEO at Lamborghini, and moved to Bologna in April 1994, successfully returning the company to profitability. Mike spent much of the next decade in the Far East. Whilst working for Tata Motors in Mumbai, advising on forward strategy, product planning and troubleshooting, in early 2005, he was invited by the Proton Chairman to join Proton's boards, one of which happened to be Lotus, and he thus became a Director of Group Lotus International Ltd, as well as ten other directorships within Proton Holdings.

After identifying the potential for Lotus, together with at least 117 Essential Key Performance Indicators (EKPIs), areas of concern that needed attention, as well as new revenue-earning opportunities, including rebuilding the activities of Lotus Engineering, Mike was appointed full-time CEO. Between 2006 and 2009, he went on to turn an £11-million loss into a £2-million operating profit. I first met him at LOG28 in Indianapolis in 2008, where he introduced the Evora to the assembled US Lotus buffs.

Mike's best-known achievement during his tenure at the helm of Lotus during this second phase was to initiate, engineer, develop and launch the Evora. The Europa and 2-Eleven were also introduced under Mike's tenure, plus extended agreements negotiated with Elon Musk to extend the build of the Tesla Roadster. The Evora was further refined over the following decade, along with its Elise and Exige siblings. "I worked very closely with Colin Chapman," he recalled, "the inspirational

Roger Becker – Vehicle Engineering Director

Roger Becker joined Lotus as an apprentice in 1966, working on the Elan assembly line at Cheshunt just before the company relocated to Hethel, but his natural driving and engineering skills came to the attention of Lotus founder and unrivalled talent-spotter Colin Chapman. As a result, Roger was quickly moved to the Vehicle Development team where he worked directly with Mike Kimberley (former Lotus CEO) on the mid-engined Lotus Europa Twin-cam – his first Lotus car development project.

Roger passed away in 2017 aged only 71, but during his 44-year Lotus career, he was responsible for the development of every single model produced by the company, including the Esprit, Excel, Elan, Elise, Exige and the new Evora. He was one of the core group who maintained the philosophies laid down by Colin Chapman, encapsulated in the famous phrase 'performance through lightness,' ensured that the essence of Lotus was at the heart and soul of all the firm's development projects and new models. As a Lotus development driver, Roger was in the frontline of the genesis of the Esprit Turbos in the early '80s, and was a key player in the successful Esprit racing program of the early 1990s. But his career wasn't spent entirely behind the scenes; his magnificent driving skills were memorably deployed for the awesome chase scenes that featured the Esprit S1 in the James Bond film, *The Spy Who Loved Me*, and he took the wheel with precision in later Hollywood blockbusters featuring the Esprit.

Roger was only made a Director under Mike Kimberley. Latterly a travelling troubleshooter and ambassador for Group Lotus, Roger created the Vehicle Engineering Division of Lotus Engineering, and was instrumental in every aspect of development of the US Federal specification Elise, a milestone model. That included negotiations for the Toyota 2ZZ engine supply that transformed the Elise's dynamics and enabled its profitable sales performance in the USA. This was an immensely significant step for Lotus, since the 1.8-litre 2ZZ Toyota unit powered the Elise R, the SC, the Series 2 Exige and the 2-Eleven roadster. The Roger Becker Elise version was fitted with Lotus' ultra-lightweight forged alloy wheels, and came complete with top-spec Sport pack and Touring pack, including air-con, while cosmetically it was available in four paint colours – Aspen White, Starlight Black, Solar Yellow and Carbon Grey. The Roger Becker signature is on the rear of the car, there's a Roger Becker numbered plaque on the dashboard, and the Lotus badges are in a distinctive monochrome scheme.

Roger Becker was Vehicle Engineering Director, and had a run of Elise and Exige special editions named after him. (Courtesy Lotus Cars)

As for the Roger Becker special edition Exige, that came with the 260PS supercharged 2ZZ 1.8-litre Toyota engine, Lotus ultra-lightweight forged alloy wheels, Performance pack, Sport pack and Touring pack, again including air-conditioning. It was also available in four paint colours: Aspen White, Starlight Black, Solar Yellow and Carbon Grey, and like the Elise, it was supplied with Roger Becker's signature on the rear of car, the Roger Becker numbered plaque on the dash, plus those monochrome Lotus badges. The end of the 2ZZ marked Roger's retirement from Lotus, and the monochrome badges symbolised the final closure of that particular phase of Lotus' history.

But the Becker family name lived on at Lotus Cars for some years: as we've already heard, Roger's son Matt was the firm's top ride and handling expert, involved in honing the Elise and Exige suspension, and, crucially, responsible for defining the sublime feel of the Lotus Evora and how it behaves on the road – plenty more to come from him. Also, Roger's daughter Kristie was a Marketing Executive in the busy Communications Department from 2007 till 2011.

---

genius who founded Lotus, and I am very privileged to have worked with him for 13 years before the tragedy of losing him. I went on to run Group Lotus until the end of 1991. While Lotus was under Proton ownership, I was invited in late 2005 to join a number of Boards, one being Group Lotus. After persuading the shareholders to allow Lotus to be rejuvenated, we created and launched the company's first new model for 13 years, the Evora, based on an entirely new platform and running gear, an achievement that I'm still extremely proud of. The effort to produce it was absolutely typical of Lotus' 'can-do' attitude, enabling all the composite strands to come together at the tight time. Meanwhile, we rebuilt the Lotus Engineering consultancy globally with EV and new technology, whilst returning the Group to financial stability. There were a number of challenges, obviously, but the cornerstone to our plan and the first full new model for over 13 years is the Evora, which forms the logical next step for owners of our current range, the Elise, Exige and Europa, and a lot of customers will welcome the opportunity to move up into the new 2+2 which embodies all the great Lotus DNA characteristics that came from Colin originally, and at the

Tony Shute – Project Eagle Platform Manager

Tony Shute was Head of Project Eagle – the codename for the Evora. He started working as a Vauxhall Motors apprentice in Luton, gained a Mechanical Engineering degree from Sheffield, and graduated from the General Motors Institute in the USA. He also spent five years working as a Tyre Development Engineer with Goodyear at their Technical Centre in Luxembourg. In 1987, Tony moved to Norfolk, and joined Lotus as a member of the Chassis, Ride and Handling Group, working on several projects. He was Project Manager for the development of the Series 1 Elise, leading the Team in defining objectives and maintaining the balance between cost and weight, bringing the Series 1 to fruition in 1995, and then onwards with the Series 2 for the European market in 2000 and a Federal Spec car for the USA in 2004. He was thus ideally placed to manage the genesis of the Evora.

Tony Shute was Head of Project Eagle, codename for the Evora implementation programme. (Courtesy Lotus Cars)

Mike Kimberley returned in 2006 to restructure the company and its manufacturing strategy, and he was an enthusiastic advocate for the new regime that placed the Evora within the ambitious plans for company growth. It was a tough task to turn the company around, to get the cost base right and take a more aggressive, proactive approach to the business, and when he returned to Lotus, Mike Kimberley asked Luke and Roger Becker what they would do if they had a quantity of corporate funds at their disposal. Their spontaneous response was that Lotus should use what it knew, and what they knew best was bonded aluminium composites. And it also needed a mid-point car, an organic evolution from the Elise. Luke expanded on the proposition: "In terms of the natural rhythm of growth, same time we incorporated the same stunning style, design and individuality – the key thing being individuality – of a Lotus." Unfortunately – and that is a gross understatement – a year later, Mike slipped and fell on an icy carpark and broke his back, invaliding him into premature retirement.

Key players

Meanwhile, Lotus expertise at that time included a core of engineers and technicians who, during the preceding decade, had brought the Elise, Exige, Europa and 2-Eleven to fruition. Key players including Tony Shute, Head of Project Eagle, Chris Dunster, Engineering Manager, Peter Wainwright, Test Development Manager, Richard Rackham, Vehicle Architect, Peter Lawton, Development Release Engineer, and Rob Savin, Product Planning Manager, had all worked on numerous Lotus projects before the Evora. Mike Kimberley valued this experience, and involved such people right from the start of the project. I'd known Luke Bennett since the mid-1990s when I did a book about the S1 Elise and he gave me a guided tour of the plant back then. As Lotus Cars' Operations Director, Luke was involved from the start of the Evora programme. He was on hand when the production question had to be: how do we get a car as compelling as we can make it into the market as fast as we can?" Two major challenges had to be faced in bringing the car to the showroom. First was the process of manufacturing the aluminium chassis. Although an evolution, the Evora wasn't just about putting a bigger engine into the Elise chassis, since Mike Kimberley's vision of a 2+2 reaching out to a wider customer base ruled out a simple cut-and-shut operation. The second production challenge was the NCAP (European New Car Assessment Programme) legislative framework. The Elise complied with previous regulations and, as an existing model, survives under that regime. But the Evora, as a new car, had to comply with NCAP regulations, which took occupant and pedestrian safety to another level.

Concept direction

By working through a series of what they call concept gateways, such as using a V6 engine rather than a turbocharged four, or factoring in the 2+2 configuration, it was possible to reach the stage of concept direction – CD – when ideas started to be formalised in renderings and in computerised simulation.

FORM CONCEPT

ELEGANT PROPORTIONS DISGUISING THE 2+2 PACKAGE

DYNAMIC VOLUMES WITH FLUID GRAPHICS

MUSCULAR YET LIGHTWEIGHT FORM LANGUAGE

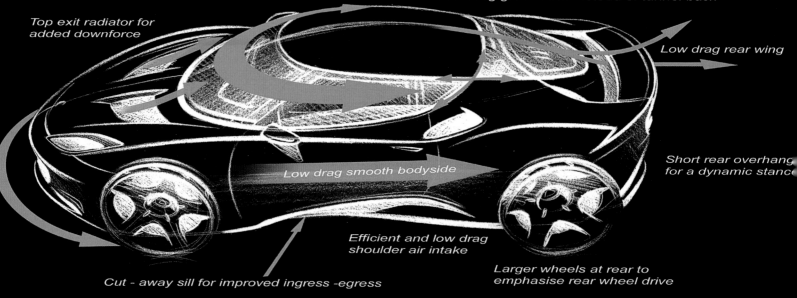

- 'Visor' windscreen + crossover window graphics to optically reduce the length and height of the cabin
- Low drag glass back instead of tunnel back
- Top exit radiator for added downforce
- Low drag rear wing
- Long front overhang sweeping back from the central intake
- Low drag smooth bodyside
- Short rear overhang for a dynamic stance
- Efficient and low drag shoulder air intake
- Cut - away sill for improved ingress -egress
- Larger wheels at rear to emphasise rear wheel drive

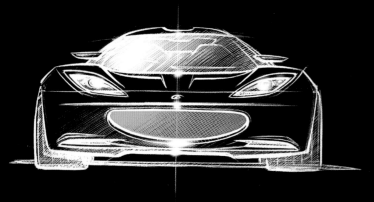

Characteristic Lotus centre air intake

Menacing yet likeable 'face'

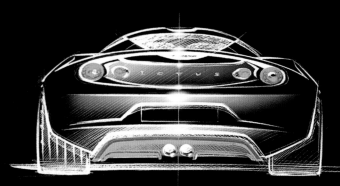

Prominent diffuser

Central exhausts

Prominent LOTUS badging

Unique rear light graphic driven by floating rear wing

Design Studio renderings detailing the form concept for the Evora. (Courtesy Lotus Cars)

I'm not that into this business-speak jargon, but all aspects of life have become encumbered with it; there's a kind of them-and-us about the use of such verbiage, and I'm on the board of my wife's charity simply to ward off the temptation to slip into business lingo. Anyway, we're rather saddled with it here, and I'd prefer to spell out the acronyms rather than saddle you with a glossary that needs constant reference!

It took another eight months to reach CA – concept approval. The Concept Approval Board consisted of just four people, and took account of submissions from the project's current and potential key agents, and from in-depth market research. Lotus commissioned a marketing agency to research a brief, looking at a basket of comparable products, including, amongst others, the Nissan 350Z, the Porsche 911 and the Jaguar XK, counting their worldwide sales, and establishing what the new Lotus' unique selling points would be against its competition. The research base provided information that helped Lotus predict what proportion of that market it would be able to attract, and therefore the number of cars that could be sold, year-on-year. With this information to hand, Lotus built up the requisite component data by consulting component suppliers on the costs of parts, so that by the time it was ready to decide, the Concept Approval Board had authoritative information in front of it on production costs, predicted sales and projected returns. As Luke Bennett recalled: "We all shook hands, and off we went! After that it was about monitoring the progress against those concepts."

The original product profile was signed off in September 2006, and the project's organisational structure was established from the outset, with the two project leaders, Luke Bennett and Chris Dunster from Engineering, reporting to Tony Shute. The Release Engineers held weekly functional meetings, attended specifically by Tony or Chris. Other key personnel brought in were Iain Snairy, Manufacturing Release Engineer; Peter Wainwright, Supply Chain Leader; John Cooper, Quality Release Engineer; Kerry Hopper from Finance, and Peter Lawton, Development Release Engineer who was in charge of the test programme. This group met once a week, juggling costs, timing and design. As Tony commented, "You can have the greatest car in the world, but if no one can afford it, what's the point?" Executive Review comprised all the Lotus Directors, plus Tony Shute, Chris Dunster, Peter Wainwright, Richard Rackham and Russell Carr. The project demanded a high level of decision-making at each stage, so referrals upwards were limited. Mike Kimberley called special meetings to review particular issues during the course of the project, but the Directors mostly took a back seat. "It's very expensive to have 20 people sitting together," declared Tony Shute, "unless they are being highly productive. But it's back to that Lotus culture: it's enabled us to cut down meeting time by being in the office, discussing the problem informally. By the time you've booked the meeting the situation is resolved, and if you need a decision quickly, your Head of Programme, your Engineering Manager, are sitting in the same office, so you just go and ask them." Today, we've become accustomed to and adapted to Zoom and, as I write this, admin staff are trickling back to the site, post-pandemic, having been working from home, though the plant has been fully operational.

Throughout the Evora programme, the project was reviewed at predetermined stages, with each group making a presentation on the status of its specific area of activity. Chris Dunster provided the release status, Rob Savin talked about automotive market-driven predictions, Peter Wainwright reported on investment in tooling and bill of materials, and Tony Shute and Kerry Hopper reported on the business case at that particular stage. The projected product profile was reviewed at every gateway meeting, though in fundamentals it changed very little since the target market had to be kept in mind all the time, along with all the characterful attributes that a Lotus needs. Occasionally, a design feature might work on the cost basis, but couldn't be justified against the schedule, which was why Lotus built the Evora prototypes on the production line in order to keep the timing on track. One hiccup in the Evora's development was the wing mirror, which illustrates the impact of a single, small, unplanned change on the whole process. As Tony Shute admitted, "We didn't get the specification correct early on in the programme, so the new, much better mirror changed the complete door design, the electrical architecture, all the switches, and that put us behind on the prototype phasing."

Engineering Manager Chris Dunster's primary responsibility on the Evora project was the release of all the production drawings, and seeing through the engineering for the first phase of prototypes. Chris joined the company in September 1969, on the same day as CEO Mike Kimberley. When the VP phase of prototypes was released for full production, he was tasked with moving the operation seamlessly through the next phases of prototypes, press cars and, ultimately, customer cars. "The prototype programme passed from EP, engineering prototypes, to VP, verification prototypes," he explained. "The EP body panels were handmade at Hethel by Lotus, whereas the VP prototype bodies were made in France by SOTIRA's RTM process."

(SOTIRA = Les Sociétés de Transformation Industrielle de Resines Armées; RTM = resin transfer moulding.)

LOTUS EVORA

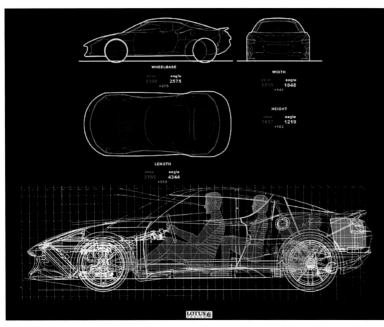

CAD renderings comparing relative sizes of Elise and Evora (née Eagle); also demonstrating the Evora could be construed as a genuine 2+2. (Courtesy Lotus Cars)

"VP testing is production-intent, because you have to prove that your test data is repeatable. Then, finally, the proper certification is obtained as a witness test. In the US, the test approval is gained through self-certification, which is in some ways more onerous, since without a third-party witness, the burden of proof and liability remains firmly with the manufacturer."

The target time-frame to creating running Evora prototypes was a mere 18 months, and was achievable because Lotus was using its own proven technologies in many cases. "The extruded aluminium chassis is one of our core competences," said Rob Savin. "We were able to produce a new version of that by applying what we'd learned since we first did it with the Elise. There were several spin-offs from that, principally the Exige, Europa, Tesla and GM VX-220, and we didn't stray far from those. Many new components were items we'd been developing for projects such as the projected Esprit replacement model, the forged aluminium wishbones for example. That was new technology for us, as far as the Lotus-badged product goes, but it's something we were familiar with from the engineering consultancy aspect. So that is partly why we could productionise the Evora in such a short period of time." It was an object lesson to the automotive industry in general.

Evora panels were made by SOTIRA in Brittany, using the RTM (resin transfer moulding) process. (Courtesy Lotus Cars)

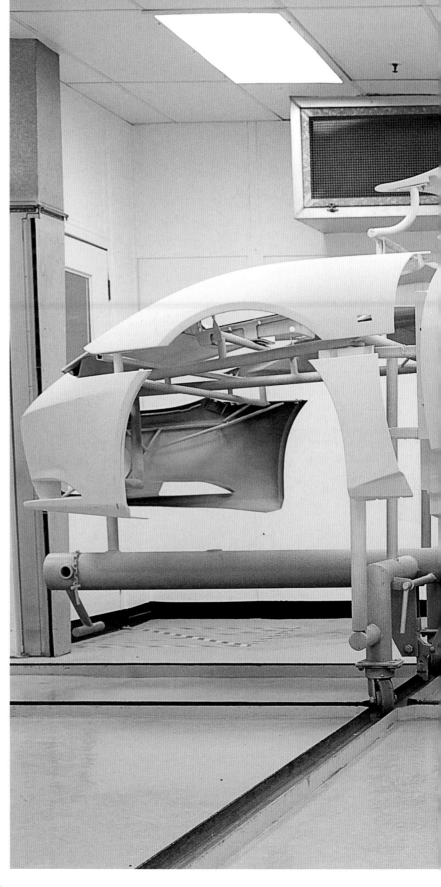

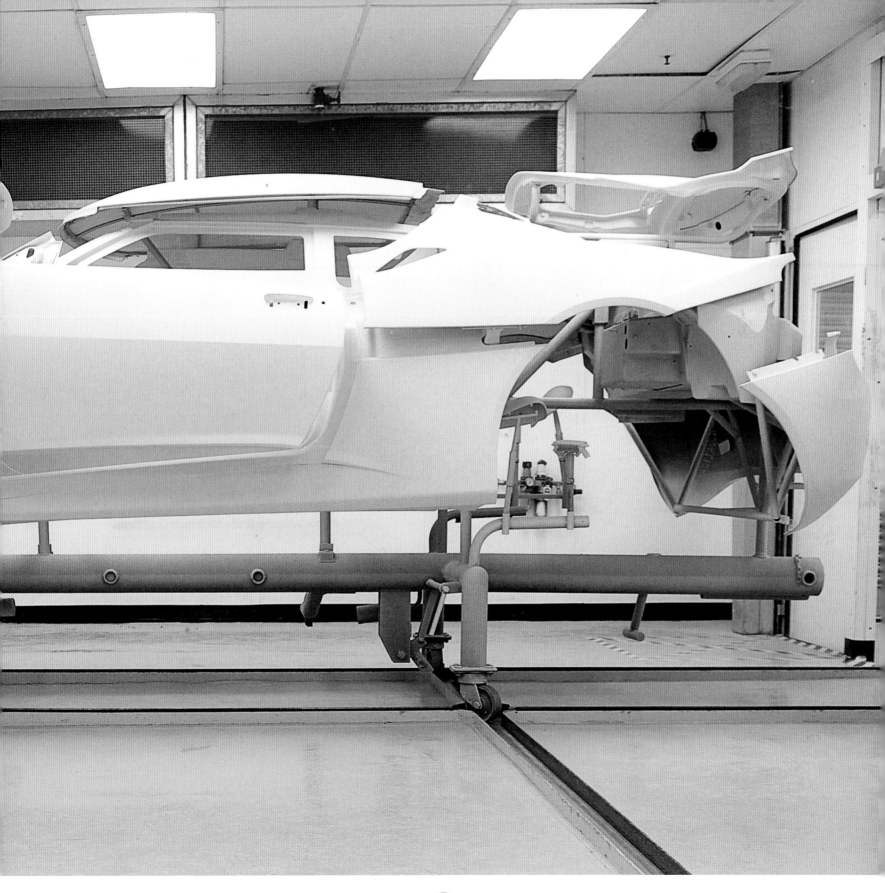

CHAPTER 2

STYLE COUNCIL
The rendition of a new model

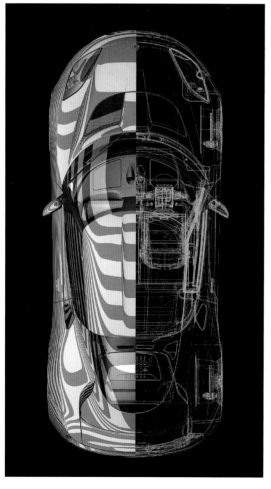

Spider's web meets mini scaffolding: overhead CAD image reveals the Evora's anatomy and form of the body shape, as evolved in the wind tunnel. (Courtesy Lotus Cars)

The Lotus Design Studio is an austere environment: clean and functional, with renderings and colour charts decking the walls, and scale models, componentry and partial full-size mock-ups randomly placed. Russell Carr is Head of Lotus Design, in charge of the team that created the look of the Evora, along with Design Manager Steve Crijns, who was responsible for the exterior of the car, and Senior Designer Anthony Bushell who was in charge of the cabin interior. While Russell was briefly overshadowed during the Bahar sojourn, he bounced back as top dog in 2014. Crijns, meanwhile, has hopped off to be senior Design Manager at McLaren – and, briefly, it looked like he would be joined by Matt Becker. Matt, however, took on the role of Technical Director at Jaguar Land Rover.

"We had a very clear, simple brief," Russell told me. "We needed to make the car visually stunning and, obviously, a sports car has to be incredibly beautiful: it has to look dynamic, it has to look modern. But, on this programme, we also had to think about not just purely form, but function as well, so we wanted to consider the car's everyday usability, which meant it had to be a 2+2. And we also wanted the design to assist in the functionality of the vehicle, so we were looking at making sure the shape optimised aerodynamics for low drag and downforce on the exterior.

"Obviously as designers we wanted to place our own standpoint on this vehicle: it was very important that it was modern, but we wanted it to have a classical twist as well, because that's what we think will last well in the market. Our

The clay models representing the final three possible iterations of the Evora, one of which got the nod. (Courtesy Lotus Cars)

Mirror image: front-three-quarter view of the three finalists for the Evora design; certain features such as the front splitter would appear in subsequent production versions. (Courtesy Lotus Cars)

cars need to look good on the day they're launched, they need to look good five years into production, and they need to look good 20 years from now, from whence they occupy the classic car market as well."

Incorporating the DNA of Lotus is an important factor as well. "We want this to be recognisable, so when we talk about Lotus DNA, a few things are relevant: first of all, the front of the car, so we have a signature grille, and we highlight certain technical features on the car and make them key characteristics of the design, so the top exit radiator, the rear wing and the rear diffuser are all there to create downforce, while other details such as the cutaway sill are very important, not least because it improves ingress and egress out of the car. Perhaps most fundamentally with any Lotus, though, is that it should look agile, athletic and sleek, and the agility aspect comes from the way we shrink-wrap the forms over the mechanical package, we tuck the body surface under, we cut away the sill as well, so the wheels look as if they're at each corner, so it has a fantastic stance, really short overhangs and no wasted space at all. The athleticism comes from the strong shoulder on the car, which you see here: if you look at the car in plan view you can see a lot of muscular shape and that is also accentuated in side view so it looks like an animal torso.

The Evora always had a distinctive front grille 'mouth,' revised quite radically with the 400 version introduced in 2015. (Courtesy Jason Parnell)

"This sleekness comes from the fluid forms on the car that run effortlessly from the front of the vehicle right the way to the back highlighted by crisp intersection lines. Of course, it's also important that we have new characteristics on the car, so we're

Interior Designer Anthony Bushell's illustrations for the Evora centre console and switchgear. (Courtesy Lotus Cars)

very proud of the visor screen which sits within this teardropped cabin, which also emphasizes the rear haunches, and it's a crossover graphic on the rear of the car where it meets the rear glass."

Perhaps the biggest surprise was on the inside: there's no hint of the minimalist austerity characterising the Elise and Exige. "Like an Elise, it is pure and sporty," said Russell, "but we have done it in a more exotic manner, more sophisticated, more elegant. Hand-crafted leather with visible stitching lines, precision-engineered metal panels with flush-mounted edges, lit switches, and at the heart of the car is the driver-focused environment. We worked very closely with Roger Becker and his team to optimise the position of the steering wheel, the gearshift, the pedals, the switches, so you feel at one with the car, and there's a modern instrument layout, with auxiliary digital dials. The surfaces flow around the cabin, echoing the shapes on the exterior, drawing your eye all around the vehicle, and closeting you inside, so you feel at one with it when you're driving it."

The brief was predictably straightforward: the Evora needed to be a beautiful sports car, dynamic and state-of-the-art modern. But the design was more than an exercise in form; it also had to be functional for everyday usability, with proper 2+2 capacity. The design of the Evora contributes to its functionality, so the shape assists the aerodynamics for low drag and downforce. "The brief was clear and simple," recalled Russell. "The integrated aerodynamics of the coupé, with its rear spoiler, were very good, so it hit all its aerodynamic targets the first time it went in the wind tunnel. The drumbeat behind the design process was about quality, and Lotus invested in the best components and the highest level of design and engineering input to create the Evora." The designers had to accommodate a comparatively large cabin for the 2+2 configuration, but, overall, the car was kept low and neat to sustain the vital positive power-to-weight ratio. As Russell mused, "Size is weight," which is pure Colin Chapman. Other classic Lotus features are the different sized front wheels, the short rear overhang and the longer front overhang. "These give the car a sense of forward movement and direction, so it's got a powerful presence on the road, which is another kind of continuity with the Elise."

Compared to the Elise, the Evora had what Russell referred to as gentler surface sculpture, and a plan-view Coke-bottle form to alleviate drag. Steve Crijns explained: "If you follow the principle of the Elise, with its Coke-bottle shape in plan,

The full-size clay buck is tested in the wind tunnel. (Courtesy Lotus Cars)

the air detaches from the bodyside and creates quite a lot of drag. The Elise has a 0.42 drag coefficient. The Evora has a 0.33 drag coefficient, which is achieved by much more gentle surface

Steve Crijns' artwork for what would become the Evora. (Courtesy Lotus Cars)

Rear three-quarter artwork for the Evora by in-house stylist Steve Crijns. (Courtesy Lotus Cars)

In production from 1970 to 1979, the Lamborghini Urraco 2.0 V8 was a mid-engined 2+2 design by Marcello Gandini. (Courtesy Johnny Tipler)

changes on the car. Some scaled-up sportscars can be over the top, disappointing and hard. We couldn't just do a bigger Elise, it would have looked completely out of scale." I'm not entirely convinced by that: look at the post-2012 longer wheelbase Exige – like the Cup 430 – and you have a perfectly proportioned scaled-up Elise. Still, he's the designer.

The science of the low-drag design, using a fastback roofline, glazed rear window over the engine bay, and teardrop cabin form, contributed to reducing drag by ensuring that the airflow stays attached, whilst the flat undertray and rear diffuser create downforce – contrary to the tunnel-back configuration, with flying buttresses that interrupt the airflow. Fortuitously, the shoulder air intake is small because it's located at the most effective position, and being minimal it presents less drag than a larger aperture.

A neat detail carried over from the M250 project was the floating rear wing, which, although integrated with the body, is a proper wing section, and very effective at creating downforce. This did create a design challenge, because air had to pass on each side of the wing, which in turn meant the tailgate surface under the aerodynamic device had to be lowered. The knock-on effect was that the depth of the rear light panel was reduced, so it was impractical to accommodate the equal-sized lamp units that were familiar on the Elise. The Evora relied more on efficient aerodynamics to provide a low drag-coefficient, which avoided the need for a really big engine that would inevitably bring a weight penalty, as well as cost more to achieve the desired speed and handling performance. Its class-leading ride and handling is partly achieved by adequate and balanced downforce. A number of styling innovations helped create downforce. At the front, the top exit radiator was a pragmatic solution for the Evora that also aided downforce, while the front splitter was inevitable in this respect. At the back end, the low drag, aerofoil-section floating rear wing was much more effective than a high drag air dam style wing. The large, zero drag rear diffuser ensured that downforce was completely balanced, front and rear.

Styling icons
Pundits in the automotive world inevitably look for influences in design and similarities in styling cues. With only the Ferrari Dino 308 GT4 by Bertone, the Ferrari Mondial by Pininfarina, the Maserati Merak by Giugiaro, and Lamborghini Uracco by Gandini to go on as examples of mid-engined 2+2s, there are not many rivals in the beauty contest. All four inevitably fall into the bracket of typical 1970s designs, but while the linear Dino and elegant Merak are tidy solutions, the Uracco seems over-long in the wheelbase, and the Mondial is an aesthetic curiosity. The Evora's frontal aspect is suggestive of its contemporary Lamborghini Murciélago and Koenigsegg, particularly in the headlight treatment, the flat planes of the front deck and swooping roofline, and from the back, the Lotus family resemblance is unmistakable. But it's back to Italy for an overall comparison: the proportions of the Evora are almost dead equal with another two-seater, the Ferrari 430 Scuderia, and that is really something, to package a 2+2 in such a fabulous shape.

In the mid-1960s, when mid-engined Le Mans cars like the Ferrari 330 P4 and Ford GT40 were taking the sports-racing world by storm, Ford thought it a good idea to style its Mustang production car with a pair of fake air intakes located ahead of its rear wheels, ostensibly serving an engine mounted amidships, despite the obvious location of its powerplant up front. Russell Carr believes a mid-engined car has to look like one: you need to know exactly where the powerhouse is. Apart from its single-seater racing cars, Lotus has been producing mid-engined sportscars since the Type 46 Europa of 1966, so no one can teach them very much. So, obviously, there's no likelihood of any detailing on the Evora being fake. Looking at the top of the car with a drone's-eye view, the teardrop shape echoed by the interior is evident as the screen wraps around the front and flows to the back. The air intakes are sited where they are out of mechanical necessity, by which they let you know that this is a mid-engined car. They work with its form, so that, looking down on the body, the real muscle curves are over the

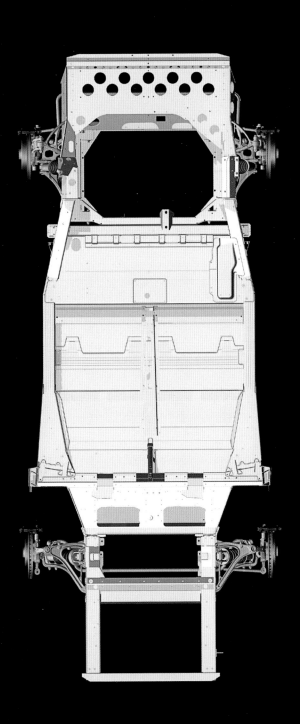
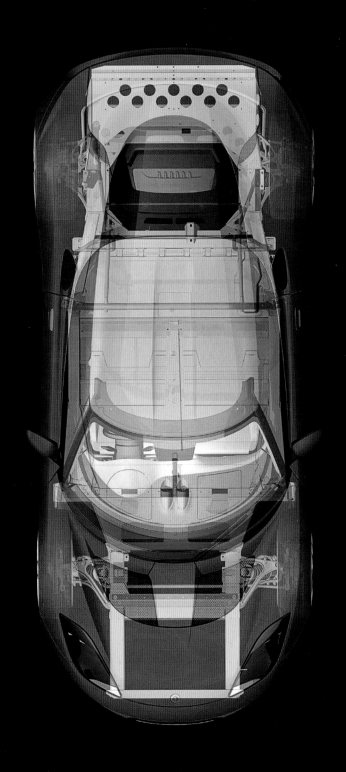

CAD images of the bare Evora chassis and with bodywork superstructure ghosted on. (Courtesy Lotus Cars)

The shoulder air intake was a feature of certain Lotus sports-racing cars of the 1960s, such as this Type 30, restored by Classic Team Lotus and driven at Snetterton by ex-F1 designer Chris Murphy. (Courtesy Johnny Tipler)

The ex-Brian Muir Willment Engineering 4.7-litre Type 30 in the Brands Hatch paddock at the Motor Show 200 meeting, 29/10/67. (Courtesy Johnny Tipler)

The Type 118 M250 from 1999 was a concept car, prefiguring the proportions of the Evora shape and V6 powerplant. (Courtesy Lotus Cars)

rear haunches. Viewed from the hindquarters, the Evora is all drama, displaying all the attitude of a lion poised to pounce. The shoulder air intake is positioned in the most efficient place to take air into the engine, which means only a small intake was needed minimizing drag. Once again, this refers back to Lotus racing cars, since in the 1960s, the Types 19, 30 and 40 sports-racers all had 'shoulder' air intakes.

Of course, it was important that new and innovative characteristics appeared in the design as well, so Russell was proud of the visor screen that sat within the teardrop cabin, once again emphasizing the rear haunches, in particular where the crossover graphic created by the 'touch-down' between the rear-quarter glass and the rear screen occurred. The visor screen was a detail developed for the Type 118 M250 project, which was retained but modernised for the Evora. The visual reference was a crash helmet, and it followed the flowing lines around the car, contributing to the sense of movement. By binding the whole design together, the visor screen also helped to reduce the visual impact of the car's extra height, tricking the eye into perceiving it lower than it really was, by playing with the proportions.

The frontal aspect of the Evora bore a family resemblance to the Elise, but also with basic functions, crash structure and radiator beneath the skin. The car was designed around the prevailing regulations, so it conformed absolutely to crash test requirements. It was also the first Lotus designed to contain pedestrian impact, so a lot of work was done to make sure there were no rigid parts on the car for head-fall impacts, which was an interesting process for the design team. There were concerns that the front splitter would perhaps create too much downforce, but it had to be there to meet pedestrian impact regulations. These stated that the first point of contact with the pedestrian must be the lowest point on the car. Because the front bumper surface was tucked under to create a lightweight looking front-end, it was necessary to create a splitter extending forward so it would be the furthest point forward. This large splitter created more downforce at the front, which then had to be balanced with more at the rear.

There were six full proposals leading up to approval, so the final car was a very condensed design. Russell talked me through some sketches from the early phases of the design programme. "There was a link with what we were doing with the Elise. We envisaged younger buyers would come to the Evora through the

The frontal aspect of the face-lifted Evora bore a family resemblance to the Elise, as in this 250 Cup version. (Courtesy Laura Hampton)

From all angles, with above being no exception, the Evora's proportions are graceful yet purposeful, and also entirely harmonious. (Courtesy Lotus Cars)

The Type 115 Elise GT1 was a broad-bodied version of the standard road car, created to race in the GT1 category. Just seven were made in 1997, plus this single road car. (Courtesy Johnny Tipler)

Elise, so this was a curvaceous form language, where we mixed soft surfaces with crisp intersection lines. The price point of the Evora meant that its design had to be more grown-up than the Elise or the Exige. Our small-platform products are much more closely related to track cars, and therefore it is only natural that they should have a slightly more aggressive and raw character. However, we were also aware that some larger sportscars often become too sterile and, ultimately, lack the passion and excitement to justify their existence. Therefore, our challenge was to create something that was both dramatic and elegant at the same time. We couldn't simply create a bigger version of the Elise, because the lines and surfaces would have looked out of place on a car with the Evoras proportions and personality." (There had been a precedent and, indeed, they did: the Elise GT1 was a wide-bodied racing version created in 1997, and the Exige S was built on an extended chassis from 2011.)

A trio of scale models was narrowed down to two, and eventually the selected design was turned into a full-size clay model. The team kept reference cars in the studio, because it's such a deceptive visual environment, and can either unfairly flatter or condemn a model. The reference cars provided a visual key for the creation of attractive proportions. Lotus turned the clay scale models around quickly for the Evora: each ranged from three to five weeks in construction, with one model dropping

Clay modeller Martin Pye scrapes the model's roof into shape. The mirror image ensures both sides match. (Courtesy Lotus Cars)

STYLE COUNCIL

Clockwise from above:

Tools of the trade: implements used by the modellers to sculpt the clay buck. (Courtesy Lotus Cars)

Carving out the clay buck to very specific measurements defined by CAD. (Courtesy Lotus Cars)

Wrapping the clay buck to create a finished appearance. (Courtesy Lotus Cars)

A few design similarities were carried over to the Evora from the Barney Hatt-styled Type 121 Europa, seen here in the Atlas Mountains on my run from Lotus' press car depot in Stuttgart to Marrakech. (Courtesy Jason Parnell)

out to be replaced by another. The selected scale model had fundamentally all the key characteristics of the final car, with the visor screen, the side view with the cutaway sill, and the Lotus mouth. The only thing that really changed was the lights. An early rendering had a pair of driving lights mounted either side of the radiator air intake, similar to the modern Type 121 Europa.

The Evora's Bi-Xenon headlamps were a major investment for Lotus, since they were designed specifically for the car. It's always tempting for smaller manufacturers – major ones, too – to avoid such outlay by using off-the-shelf designs; for example, the Morgan Aero 8 used Mini lights. The Evora design called for headlights that made legal visibility angles difficult to meet, because the beams needed to be visible from a 45-degree angle inwards. The decision was made to use Bi-Xenon lights for optimum illumination, and to use LED indicators to ensure the investment was at least providing cutting-edge technology that was going to stand the test of time. The registration numberplate obscured the front intake slightly, but with the low front there was really no other place the designers could put it. In the USA, some states don't require a front plate, which is great, but it's a fact of (ungainly) life in the UK and most other countries. There was a time when stick-on plates were fashionable, especially when road-registered race cars were driven to circuits.

Cabin fever
The Evora was designed to project all the drama and visual excitement of the Elise and Exige, whilst being more useable. It would still be more hardcore than a standard Porsche 911 or Cayman, but softer than an Elise. The key factor to defining that difference was the ease of getting in and out of the cabin, so the design detail of the cutaway sill was crucial. The sills were lowered, and, because the car was taller, the door openings could be taller. The seats are about 65mm higher than the Elise, relative to the floor panel, so you slide rather than drop into the car, and that makes access much easier. CEO Mike Kimberley could sit in the Evora, and he was 6ft 5in. As a dedicated hat-wearer and six-footer, I can sit in the Evora with hat on and still have a decent amount of headroom. There was a time when supercars were designed solely for looks, and anybody taller than 5ft 8in probably wouldn't fit. Today, of course, this is unacceptable, as average heights have increased and taller people have to be accommodated.

Getting this right was a design goal for the team, because they managed to integrate form and function perfectly. From the function point of view, pushing the sill in close to the structure meant that it was possible to minimise the amount of bodywork the passenger had to climb over, so that, when exiting the car, occupants throwing out their leg effectively placed it below the

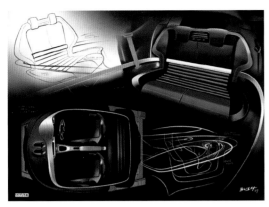

Renderings by in-house designer Anthony Bushell for the Evora cabin interior. (Courtesy Lotus Cars)

CEO Mike Kimberley's height – 6ft 5in – had a specific bearing on the proportions of the Evora cabin. (Courtesy Lotus Cars)

door opening. Russell was upbeat. "It's a cool visual piece when you look at it from rear three-quarters. The upper part of the body side can be a little gentler to reduce drag, but down there where it's not an issue for drag, you can put a lot of sculpture in. It's reminiscent of racing cars where they vent the air out, and that line works with the top line to accentuate the feline haunches over the rear wheel. It's got a much more three-dimensional shape in plan-view, which no one else really does."

Both the 2+0 and the 2+2 have the luggage boot in the pit behind the engine but, whereas the 2+2 configuration majors on its rear seating, the +0 car provides space instead for full-size suitcases behind the main seats, so it is still a practical proposition for shopping, accommodating a dog, or elements of the notional drum-kit. The +0 and the 2+2 have subtly different interiors, though most of the cabin parts are shared.

The inside of the Evora was, up to a point, reminiscent of the Type 121 Europa,

The cockpit of the Type 121 Europa coupé was essentially a two-seater, though far more palatial than the austere Elise. (Courtesy Jason Parnell)

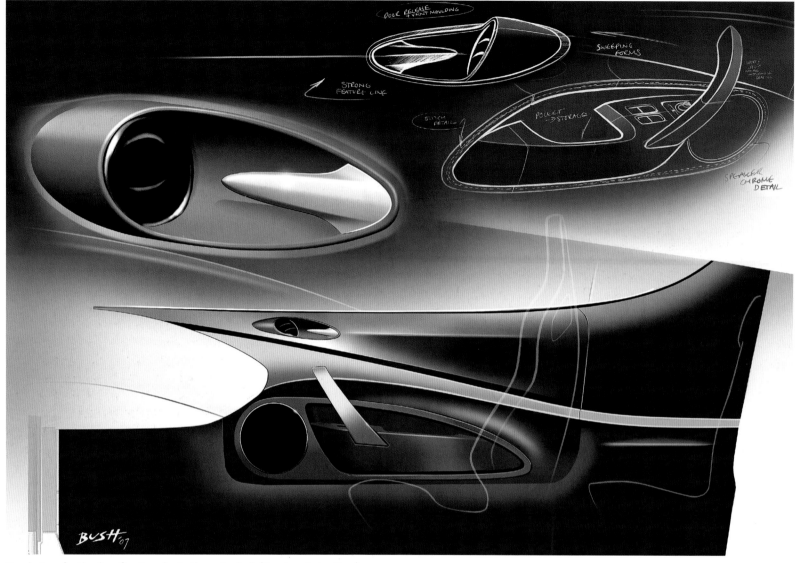

Renderings for the door furniture by Anthony Bushell. (Courtesy Lotus Cars)

the plusher GT version of the Elise, in so far as it was pure and sporty, but the design was more exotic, sophisticated and elegant, less beholden to the smaller car's raw cockpit architecture. The upholstery was handcrafted leather with visible stitch lines, and the panels were precision-engineered metal with flush-mounted edges. The heart of the car was a driver-focused environment, so Russell and Lead Interior Designer Anthony Bushell worked closely with the engineers, in consort with Roger Becker, to optimise the position of the main controls, including the steering wheel, gearshift and pedals, so the driver would feel at one with the car. From the driving position, the state-of-the-art instrument layout was clearly laid out in a binnacle that looked like a Starship Trooper's goggles, with auxiliary digital dials. The UK-sourced bespoke instrument cluster included tyre pressure monitoring, door opening, a computer, satellite navigation and reversing bleepers. There was also a reversing camera to help with rearward vision, its imagery displayed on the satnav screen. Stylish momentary switches required minimal pressure and were illuminated by a circle so they were visible as soon as they were turned on. The surfaces flowed around the cabin echoing the shapes of the exterior, drawing the eye around the vehicle and closeting the occupants inside, so that there was a harmonious experience between car and driver.

According to Russell, "The Evora was a huge departure from what people might have been expecting, based on Lotus' recent past. The flat-bottomed steering wheel, figure-hugging

Not only the exterior form of the Evora was modelled in clay, the cabin interior was similarly sculpted as a life-size cockpit environment. (Courtesy Lotus Cars)

sports seats, contemporary instrumentation, and ergonomically positioned controls provide an intuitive environment that forms an instant connection between driver and car, ensuring that it becomes an extension of his or her body."

Senior Designer Anthony Bushell explained the choice of materials in the Evora's interior: "Tactile quality is extremely important within this segment, so for the Evora interior we selected honest, premium materials. The instrumentation is unique to the Evora, and much of the switchgear is bespoke, every metallic component is, actually, metal, and leather comes from a natural source, hand trimmed to ensure high craftsmanship."

When conceiving the cabin interior, Anthony started off with the buck for the cockpit, built up according to his instructions by the in-house clay modellers. "We started off with a set package within the product profile of what needed to go into the interior," he explained. "We had to work very cleverly to package items, such as storage areas in the doors, glovebox, armrests, all of which were in the spec from the start, but to maintain an interior that looked and felt lightweight as well as coherent. So, given all of these constraints, trying to create a dynamic environmental form was the hardest challenge that faced us."

It was crucial that the cabin projected its 2+2 ergonomics, yet still had the positive feel of a driver-focused cabin. One major, and potentially problematic, addition to the cabin spec was the satnav unit: a relatively large volume within the dashboard. "It was a boxy unit, so to create a wraparound dynamic form but still incorporating that box was really difficult. The introduction of horizontal lines that swept seamlessly into the doors takes the eye away from that volume in the centre. The satnav unit forms part of the driver-focused controls that incorporate the instrumentation, which are a nice mix of analogue and digital displays. These instruments have a contemporary feel, framed by a chrome finisher, whose unique shape has become a major

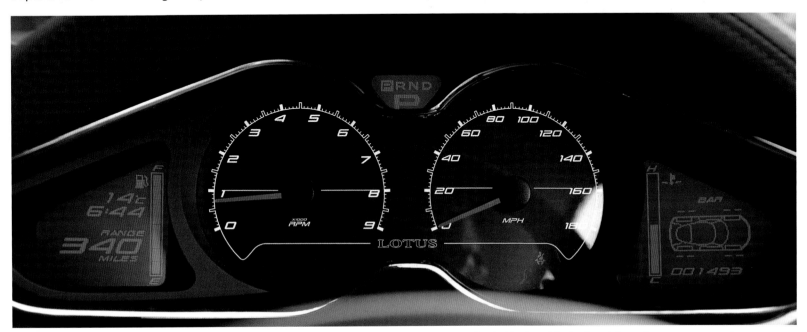

The Evora's main dials are housed in an attractive binnacle, with analogue speedo and rev-counter flanked by digital fuel info at left, and tyre pressures and temperature at right. (Courtesy Lotus Cars)

LOTUS EVORA

Steering-wheel controls, dashboard dials and switchgear became more comprehensive and business-like with the 2017 GT430. (Courtesy Lotus Cars)

The Evora presents a perfect driving position, as I discovered on an early drive in a press car. (Courtesy Michael Baillie)

Evora drivers are well catered for in terms of instrumentation: the satnav screen doubles as a reversing display. (Courtesy Lotus Cars)

showpiece of the interior. Being a driver-focused car, the instrumentation and switchgear form the central element from which the surfaces shoot off around the cabin."

It's one thing designing a beautifully refined cabin with sweeping forms, but it's quite another ensuring that all the instruments and controls are located exactly in the right places. To achieve this, Anthony worked with Matt Becker and the driving dynamics people to ensure these design showpieces were functionally correct and accurately sited. "All the switches, stalks and instruments went through many iterations to get them absolutely right and in the optimum locations."

As with any car, the steering wheel is another key element to the driving experience, as Anthony explained: "The steering column itself was a carry-over item, but the steering wheel was a unique design to the Evora; we put a flat bottom on the rim, which looks fantastic and also helps ingress and egress to the cabin as well. We knew from the outset that we would be using Recaro seats, so we designed those into the package too." The clever detailing within the design helped generate a premium feel. For example, the contemporary door releases are distinctive, but feel high-quality, as do the air vents. "We put the vents on the door to achieve more space within the instrument panel and, by doing this, we created a dramatic feature. The vent moulding is actually the same part, just flipped from side to side, which is a useful reduction in tooling costs."

Other visual devices influenced perception of the cabin environment. "Inevitably, the exterior design had a big influence on the interior. When sitting in the car, having such a dominant visor-shaped windscreen ahead of you, the interior needed to be a strong, embracing environment, and we built upon this to provide a fresh, exciting feel to the cabin. When you sit in the car I want you to feel the drama of these wraparound forms working together with the screen to create a cocooning cockpit, you need to feel that you're a part of the driving machine. Therefore, instead of having separate sections, like a definite instrument panel and door, you have an entire surface, a swathe effortlessly sweeping around the cabin." These interior forms harmonised the exterior and interior, and the surface language had similar qualities.

Outside, there were soft forms with crisp intersections, and those were reflected in the interior, like the crisp intersections on the centre console and the upper door panel. The different colour bands and clever trimming methods created floating elements that broke up the masses and presented a lightweight look to the cabin. The colourway band ran around the cabin and linked into the rear seating area, framing the environment and its occupants. The front seats were trimmed and of course styled in a different upholstery to the rear ones, emphasising the point that this was very much a drivers' car. Cramped as it might be, the well-being of potential rear seat occupants was not ignored, nonetheless: "We needed to make it look welcoming in the back, because a lot of 2+2 grand tourers just offer a simple bench seat. We worked with all these dynamic forms, and the 2+2 seats reflected that." The levels of craftsmanship in the fully leather trimmed interior were impressive. "We wanted to make a point of creating surfaces where we could sew dominant stitches into the leather, because you can have surfaces that are leather-trimmed, but it's the runs of stitches in front of the driver that reinforce the quality of craftsmanship that has gone into the design."

In the Premium Pack that was one of three optional packages offered in the launch models, focused courtesy lighting was installed throughout the interior, generating a spectacular illuminated feel to the cabin's design features. The extensive use of handcrafted, premium leather differentiated the Evora from other Lotus products, and such was its importance to the character of the car that an entirely new trim shop was created within the factory, dedicated to its production. Even the entry-level versions of the Evora were fitted with leather seats, together

Visitors to the Lotus stand when the Evora was unveiled at the London Motor Show were shown the range of available exterior colours and upholstery shades. (Courtesy Lotus Cars)

The Evora GT430's rear end is just as purposeful-looking as the front, featuring bridge spoiler, cooling vents and carbon-fibre diffuser. (Courtesy Lotus Cars)

with upper door trim panels and facia upholstered in leather, it was anticipated that the majority of customers would go for the full hide treatment and accent lighting available in the Premium Pack. Along with four different shades of leather, available in the Premium Pack, customers could equally well personalise their Evora with a broad range of paint hues.

The Lotus colour palette featured 20 colours in 2008, which was the most that the paint shop could cope with at the time, considering they had to work in all markets and with all the other models in the Lotus range. Until Evora came along, the colour range had only to cater for the Elise, Exige and Type 121 Europa, and included a number of loud colours that suited those models very well. The Evora, being a more 'grown-up' car, was thought likely to be ordered in less in-your-face hues. Several colours were deleted from the range and replaced with new ones specifically with the Evora in mind: Quartz Silver, the colour the Evora was launched in. Silver is considered the best colour to show off the shape of a car, but it is a tad dull and conformist. Many Porsche models are traditionally finished in silver, suppressing potential excitement over performance in the interests of conformity and modesty – and the duller for

it. But that's more or less why Lotus developed these colours – Quartz, a warm silver with some depth in it, Aquamarine Blue, a sophisticated blue green metallic; Carbon Grey, a warm dark grey metallic, favourite in the Design Department. Said Anthony, "It manages to make the car look expensive and sophisticated, as well as mean and moody and, although it is dark, it still shows off the window graphics quite well. On top of that, customers could choose between gloss silver, cast or forged wheels, matte Stealth Grey cast wheels, or gloss Anthracite forged wheels. The silver wheels gave the car a sophisticated feel, the dark wheels created a sportier look, and complemented bright-coloured cars such as yellow and white."

The rear diffuser, twin tailpipes and minimal light clusters set off the rear aspect of the car, complemented by the shallow arcing spoiler bridging the sides. In the spirit of Lotus' philosophy and DNA, lots of technical detail was built in, played up by the designers to support the powerful visual sensation of the whole car. Everything was in place for a reason: the diffuser provided downforce, as did the top exit radiator, and the intake cooled the engine.

Working alongside the Design team, Richard Hill, Lotus'

STYLE COUNCIL

Rimstock was the supplier of the original ten-spoke Evora wheels. (Courtesy Lotus Cars)

The Evora 410 Sport's ten-spoke wheel is shod with Michelin Pilot Sport Cup 2 tyres. (Courtesy Lotus Cars)

Vehicle Development and Homologation Director, was responsible for fine-tuning and signing off the Evora's aerodynamics. "Ultimately, a sports car has to be a balance between aesthetics and function," said Russell. "Richard knew that if he treated it as an Imperial College project he would get great figures, but the car would not sell. But, equally, we knew that if the design team ignored the basic rules of aerodynamics we would end up with a car that wasn't stable, or overheated. So, we were sure to attend the wind tunnel sessions so he didn't hit the car with the ugly stick when we are not looking!' Bearing in mind that the Evora is one of the most graceful, yet purposeful-looking sports GTs ever made, it was obviously a synergy made in heaven.

Rimstock wheels
The Lotus Design Studio also contained stacks of wheels, important attributes not merely as tyre carriers but enhancing the styling and appearance of the car. Configured to be as lightweight, durable and efficient as possible, several designs were tried by Lotus, including examples from manufacturer Rimstock.

Rimstock had previous form as the supplier for the Elise, so Lotus knew Rimstock would pull out all the stops to get the wheels right from the off, based on its grasp of Lotus technical design standards on a low-slung car. With a low centre of gravity, the suspension loads up the wheels, but the Rimstock wheels passed all the bench tests with flying colours.

Wheels could not be designed solely on computer; they had to be validated in three dimensions, and there were always things that couldn't be seen on the screen. In the initial production process, Lotus sketched the wheel and then produced a computer model. This was sent to Rimstock, which ran the model through its software. It then modified the design to technical criteria before machining one wheel. Lotus checked the result, and worked with Rimstock to realise further modifications, and then Rimstock forged a prototype, and ultimately that became the production wheel.

The Evora's wheels were designed to look right up to date, and to help control the visual mass of the car. Big wheels and arches restyled the body to make it look lower that it might

Evora wheel design evolved along with the car; this is a ten-spoke alloy on a 410GT. (Courtesy Jason Parnell)

otherwise do. It was a common trend, because with people getting bigger and cars getting taller, so manufacturers were upscaling wheels to balance out the size of the car.

Rimstock made and supplied wheels to car and bike manufacturers worldwide, such as Aston Martin, Jaguar and Triumph Motorcycles. It operated from four factories in the UK's West Midlands, and its best-known aftermarket alloy wheel brands were Team Dynamics and 100+. Rimstock also developed an ultra-lightweight range of cast, heat-treated and forged aluminium wheels, Pro-Race and Pro-Forged, designed for competition purposes, and manufactured Sport Dynamics wheels for off-roaders.

A new option appeared in 2010, in the shape of larger diamond-turned wheels, up from 18/19in front/rear to 19/20in, fitted with the latest generation Pirelli P Zero Corsa tyres, with a more conventional tread pattern, making them less noisy and more grippy in the wet.

A MODERNIST ADVENTURE
An appraisal of the Evora by design guru Stephen Bayley

Architecture, automobile, or artefact – if it's been designed, Stephen Bayley will have a view. And he won't spare the creator's blushes. How, I wondered, would he rate the Lotus Evora?

Co-founder of the Design Museum Stephen Bayley with an Evora near his south London home. (Courtesy Michael Baillie)

STYLE COUNCIL

Design guru Stephen Bayley publishes his critiques in leading media worldwide, sparing nobody's blushes. (Courtesy Michael Baillie)

No obsequious platitudes: even if it's only slightly compromised, Stephen Bayley will have something critical to say about it. The design guru publishes his opinions in leading media worldwide, and if something affronts him, he rips it to shreds. Eloquently, mind – he's no headbanger. Bayley deploys his judgement as skilfully and as ruthlessly as a surgeon's scalpel to dissect his victims, the aberrant designer and his abomination hung deftly out to dry.

I'm confident that he'll give the thumbs up to the Evora so, nothing daunted, I park one outside his South London office for his appraisal. Trendily suave and film-star dashing, it's hard to credit that this amiable 50-something would just as likely administer a savage tongue-lashing or caustic critique. Instead the tone is complementary and simpatico. His wife, Flo, makes cappuccinos while we chat on the patio.

A graduate of Manchester University and Liverpool School of Architecture, Stephen lectured in Art History at the University of Kent in the '70s. He set up the Boilerhouse Project at the V & A Museum with Sir Terence Conran in the 1980s, presenting 20 major exhibitions from cars to hi-fi to fashion, and he went on to found the Design Museum in 1989. Then promptly resigned. He's since written eight books and innumerable articles, lectured around the world and is a consultant to a number of top global brands.

Though a car connoisseur, Stephen is also a pragmatist. A city man to the core, he was at the time driving a Smart and his wife a 10-year-old Mégane Scenic. He's also a prophet of doom, believing that the hands of the clock are at five-to-midnight for the petrol-powered motor vehicle. "Not this year, not next year, but soon we won't have cars as we know them anymore, we will have hybrids, and electric cars will become more and more efficient. So, we will save the world by making it more boring!"

And that's not a word you'd readily associate him with. One of his books from 1986 was titled *Sex, Drink and Fast Cars*, "I'd been writing books about design for years and my publisher said, 'What are you really interested in?' So, I just said, 'Since you ask, sex, drink and fast cars,' and he said 'That is a fantastic idea, do it.' So I did, and it's become a sort of classic in arts schools because it's a complete distillation of movies, rock music and what cars mean to people."

Talk turns to his latest book. Why is there only one Lotus in it – the Elite? "In terms of what I think are the all-time greatest designs, my criteria wasn't technical innovation, certainly nothing to do with performance, but cars that somehow or other set a new standard. According to that very stern criteria, there were only 80 or so cars which are absolutely radical and change the way we see things, so given the tiny size of Lotus in the grand scheme of things, even one is really quite extraordinary."

He makes the point that, "the Elite has very interesting technology, and in the history of the motor car interesting technology often leads to interesting design." Then comes the revelation. "It's honestly true to say that my introduction to car design was actually a Lotus. My father worked in the aircraft industry and always had nice cars so I've always been familiar with machinery. I remember as a boy being driven past Aintree Racing Circuit and I saw a Lotus Eleven on a trailer. It was dark blue with red Wobbly-Web wheels, and I remember thinking, 'How can anything be so perfectly, exquisitely beautiful as the Lotus Eleven,' and that was the germ that kindled my interest in design."

The sleek Lotus Eleven with Wobbly Web wheels kindled Stephen Bayley's interest in design. (Courtesy Johnny Tipler)

Which raises the question, why isn't it in the book, then? "Too much of a competition car," is the answer, though in fact Stephen's earliest foray into journalism featured another Lotus racer, the Lotus 49. "The very first article I published was when I was 15 in 1967. I had just taken some photographs of the new Lotus 49 F1 car's suspension and I did some immaculate schoolboy drawings from them. They were published in a magazine called *Model Cars*." He's never been bitten by the motor racing bug though. "There are a lot of things I disapprove of which, perversely, I find fascinating, and motor racing is one of them. But like the Lotus Eleven, you look at the Lotus 49 and you think, my god, how did they think of this? It just looks fabulous, the suspension, the simplicity, the sheer beauty of those wheels. Then there is that Chapman innovation of bolting the engine onto the back of the chassis. It was such an exquisite thing and of course it won first time out at the Dutch Grand Prix in 1967." And that historical gem is delivered by a man who claims he doesn't like motorsport; he sure wears his learning lightly.

Stephen regrets never having met Colin Chapman. "Clearly, he had extraordinary design and engineering talent himself, but even more important than that – and this is the paradigm of the whole design and creative type – he was able to motivate others, even other great designers, which is what all great designers actually do. Like Harley Earl at General Motors, they don't necessarily do it themselves, but they have the ideas and what defines their extraordinary creativity is getting other people to realise their ideas. In Chapman's case that was Frank Costin with the Eleven, Peter Kirwan-Taylor with the Elite, and then later on Len Terry and Maurice Phillippe. It fascinates me that he was brilliant, intuitive, ideas of his own, but he was also such a powerful personality."

He has not missed much of Lotus' history. "One of my absolute favourite cars of all time is the Lotus 23 – so aesthetically pleasing, and of course the wedge-shaped Type 72 with the side radiators." Before its quintessential black-and-gold JPS period the Type 72 was liveried in Gold Leaf colours (from 1970 to 1971), prompting the question, to what extent does the livery of the car effect its design? "There is a profound but not very well understood relationship between colour and perception as far as aesthetics are concerned. As a general rule, red and white don't work very well on large cars, but no one really understands why; it is an interesting area of perceptual psychology." And that thought diverts Bayley's flow into Chapman's talent for branding. "Lotus livery was wonderful: another favourite car that featured in my huge design encyclopaedia from 1985 was the Lotus 25/33; if the 49 isn't the greatest racing car ever, then the 33 must be, fabulous in green with the yellow stripe.

As a schoolboy, Stephen Bayley made drawings of the Type 49's complex suspension setup. (Courtesy Johnny Tipler)

"One of Chapman's many innovations was what we now call corporate identity. The expression wasn't known in 1970. But first of all, to call it Team Lotus instead of the Lotus Racing Team, is in itself incredibly clever coinage. You would have to pay an agency £5 million in fees to come up with something even as half as good nowadays. We tend to forget now how clever Chapman was, introducing the whole branding business into F1 with Gold Leaf Team Lotus."

The Design Museum on London's South Bank has hosted automobile-focussed exhibitions from time to time, and 20 years ago when Stephen was its first Director, he wanted to stage a show specifically about Lotus as a design object. To this end he visited Hazel Chapman at East Carleton Manor, and though nothing ever came of it, the marque has often bubbled to the surface of his consciousness. "For me the last pure road-going Lotus was the Elan," he reveals. So why didn't that make the

STYLE COUNCIL

A meeting of minds: Tipler and Bayley concur on the Evora's design attributes – though Bayley questions its suitability in London. (Courtesy Michael Baillie)

book? "Ah, but you couldn't have two Lotuses as that would have been disproportionate. I genuinely tried to get down to the essence because I think it is the end of the motor car; I don't say this with any pleasure but the great age of the motor car is over. That's really what my book is about. When we look back and ask the question 'What was this extraordinary design adventure represented by the automobile?' I wanted this book to answer, 'This is it, this is the motor car, its design, its architecture, in aesthetic terms an amazing achievement.' So as there are only 80 cars in the book you couldn't have two Lotuses, it would have been disproportionate. Although the Elan is very beautiful and important as an icon of the '60s, I think on balance the Elite was more of a step up, more of a category change than the Elan was, though in fact either could have been chosen."

A quick tally shows Renault has the most entries. "Very often in the automobile industry, economy cars are the most technically interesting because it requires great ingenuity to make sense of a packaging requirement. But equally economy cars are seldom beautiful because of their proportions. It's very difficult to make a small thing beautiful, but the same applies to racing cars. The constraints of the racing car stimulate genius. Lotus never designed a large car, and I suspect if they had done it wouldn't have been anything like as lovely as their sports racing cars because the disciplines aren't there."

Cars that are easy on the eye are usually sports or grand tourers, though as he says, there's often a price to pay. "The proportions you get out of a mid-engined layout lead to ergonomic calamities. Visibility is a terribly important performance asset, so driving Lamborghinis and things like that must be a total nightmare: yes of course it goes very quickly but if you can't move out safely it doesn't strike me as high-performance. The Ferrari Mondial 2+2 was surely ugliest Ferrari ever, which is something to be able to say – who'd have thought of an 'ugly' Ferrari. The proportions just don't work. The Evora is immensely clever, but I do think it is quite extraordinary that Lotus has got commitment and enthusiasm to do something like the Evora at this very late stage in the history of the car."

So, does he judge everything his eye alights on? "I have a completely aesthetic view of the world; the vision is overwhelmingly important to me. The whole Modernist adventure can be summarised as a desire to tidy up, and I want to tidy up anything I see, to create order. For me design isn't just that two per cent of fashionable activity which makes you look funky, for me the whole world is a design matter – anything which is made has value and meaning but some things have better value and more meaning than others. That is what interests me. If you grew up somewhere like Liverpool when I did, you certainly saw a lot of privation, and I used to see one arrangement of stone, steel, bricks and glass which was utterly disgusting but then another arrangement of stone, steel, bricks and glass could be unbelievably beautiful and exhilarating. The difference is that one of them was designed intelligently and the other one wasn't. The architect Le Corbusier said design is intelligence made visible, which is good enough for me. Intelligence takes different forms; it would be hard to argue that the 1965 Pontiac GTO which is in my book is a particularly intelligent car, but its connection with music is enough to justify its place in the history of vehicle design. That is all you can say about America, isn't it? Somebody sang a song about the GTO (the Beach Boys) and they were thinking about a Pontiac, not about Lancia or Ferrari."

He has a particular take on the future of motoring. "Like 80-per cent of people I'm concerned about the environment, but the truth is, any powered vehicle is destructive of the environment: you just have to choose what bit of the environment and what part of the cycle you want to be destroyed. I would imagine no one has ever scrapped a Lotus, ever. Some have been written-off but once they are made they last forever, so frankly what comes out of the exhaust is trivial, that is what annoys me about the environmental thing. Yes, I want everything to be cleaner and more efficient but the Prius is complete nonsense, apart from in London because you get the congestion charge free, but you are not saving the planet. I would quite like to try a Tesla, and I think electric cars will save the planet, but we will all die of boredom."

"Will a new design language emerge to define the electric vehicle?" I ask. Stephen doubts it, though he says leading car designers are on the case, even if some ideas are absurd. "Patrick Le Quément of Renault was telling me they can set up an electric motor to behave like a petrol engine; they can even set up the sound as well. But there is something for me between the noise of the engine and the idiosyncratic design of the car. I just don't think you can start fiddling with responses like that."

The conversation took an existential turn. "It's all the more beautiful that something like the Evora exists if you share my view that this is the twilight of the motorcar. It is the melancholy of this lovely looking thing, which is a delight to be in and lovely to drive, that's part of the magic of the motorcar, that against all logic they still have that ability to take us places, not from A to B but on internal journeys rather than external. The motorcar still retains the power to be emotive transport. General Motors' cars of the Harley Earl era (1940s and '50s) were of no great interest technologically, but in terms of consumer psychology they understood everything. Harley Earl said, 'You can design a car so that when you get in it, it is like going on vacation for a while,' and that is what all great cars have, even the cheap ones. You get into a Renault 4 and you think, hey, I am in French countryside, la France profonde; no doubt you get into your new Lotus and you think, wow, I am batting across Andalusia. That is what transport is, it's the internal journey."

His own journeys have been conducted in vehicles as disparate as a MG Midget and a Volkswagen Corrado. "When I was 18 my father bought me a Mark 1 Cortina, the one with the same rear lights as the Lola GT. Later, when I was a very poor university teacher a Citroën Dyane was all I could afford, but I drove it around Europe. Then I went upmarket to a Citroën GS Club Estate, one of the best cars ever, I think. I don't know why I didn't put one in the book, except there would have been too many Citroëns. I've had five Audi Quattros and five Ford Explorers and I am now on my third Smart. When I met my wife, she had a Morris Minor, then that became a black Saab Turbo. The only classic car we had was an original Fiat 500 Cinquecento. There was a marvellous moment in the '80s when outside our house we had a black Audi Quattro, a black Saab turbo and a black classic Cinquecento with whitewall tyres. With children, the Saab turbo became a seven-seater Mercedes, which then became the Renault Scenic."

The likelihood is they were never thrashed. "I am always totally at odds with the people on *Car* magazine, which I have been writing for forever and I love them all, because they only ever drive in the countryside and I only ever drive in London. And frankly, I don't enjoy driving quickly, though it is quite fun to

STYLE COUNCIL

Car connoisseur Bayley familiarises himself with the Evora's controls and switchgear. (Courtesy Michael Baillie)

drive in a spirited way, but you just can't do that in town. All car magazines write as if the four-wheel-drift is the be-all and end-all, but my argument is that you can be very enthusiastic about cars, as I am, without thinking that speed and G-force are the only criteria by which pleasure in a car can be judged. Despite all sorts of fantastic deficiencies, the Smart is an absolute delight. In London you would have to be mad to have anything else. The Smart reminds me of an old-fashioned sports car, it requires some skill to get the best out of it and you can actually drive it almost flat out."

We saunter over to the Evora. Stephen is going to take it across town to his Soho office to sample the experience. "... Low profile tyres and big wheels," he shakes his head. "... Car

Stephen Bayley's take on the Evora is, "If you accept that beauty is separate from function, you have a beautiful car." (Courtesy Michael Baillie)

An Evora in Évora: fronting the remains of the Portuguese city's 1st century Roman temple of Diana in the heart of the Alentejan capital. (Courtesy Rui Coelho)

designers must all live in the country because in London you can't avoid kerbing your wheels at some point." Let's not call up the devil ...

"My first impression is one of physical quality, and having known some olden day Lotuses, it's astonishingly impressive. But here on Nine Elms Lane in Battersea is not exactly the place to explore the vehicle's potential. In any case I don't do driving impressions because for me the motor car is more a philosophical thing. I could tell you the steering is extraordinarily quick and nicely weighted, lovely driving position, the engine revs beautifully, but we are in third gear doing 24mph, which is flat out for London these days."

Final verdict from the Design Guru? It comes with incredible fluency, and it's clever and comprehensive to boot. "To me, cars are architecture. So, the Lotus Evora reminds me of that tiny Portuguese city, the capital of the Alentejo. It is handsome, fascinating, carries a weight of history and does not fully participate in the perplexities of the modern world. Therein, of course, lies the charm of both city and car. With Lotus, the mid-engined format goes back to the '60s with the Europa, although the Evora is more directly descended from the Esprit of the mid-70s. With the Esprit, Giugiaro achieved his effects by drama rather than subtlety. Aesthetically speaking, the Evora is much more highly evolved, although it still suffers from the compromises required by a mid-engine layout. It's as if designers enter a Faustian pact. 'Yes,' Mephistopheles says, 'you can have a mid-engine and enjoy the opportunities it provides of drawing a dramatic, low shape with elegant proportions and sculptural flair.' But then he adds, 'To pay for this you will be responsible for a car with awkward access, no luggage space and terrible visibility.' Customers will be undeterred: after all, the optimal functionalist architecture for a car is the small MPV, but no one knows how to make a Vauxhall Zafira beautiful. With the Evora, if you accept that beauty is separate from function, you have a beautiful car. Even in Chelsea, it turned heads. Perhaps it would be more true to say 'especially' in Chelsea it turned heads. In this context, it would be very low church to ask, 'What is the point of this car?'"

We can safely assume he liked it, then, aesthetically and maybe even philosophically. But being pretty high church myself – as a kid, at least – I know exactly what the point of the Evora is, and with that, I headed swiftly for the open road.

Russell Carr – Head of Design

Russell Carr was a graduate of the Coventry University BA course in Industrial Design (transportation). Russell arrived at Lotus in 1990 to join Julian Thomson's Design team from MGA Developments, a UK-based automotive design consultancy, and took over as Chief Designer in 1998. Since 1990, the Lotus in-house design capacity steadily expanded from five to 30 people, and the department now encompasses design, modelling, studio engineering and digital surface development. Typically, the Design team is working on at least two major programmes, whilst simultaneously inputting to various smaller projects.

In his earlier career he was Lead Designer on the Esprit S4/S4s the Elise GT1 and was fortunate to work on some aesthetic aspects of Lotus' Formula 1 cars. Significantly, Russell led the Design Department through three generations of Elise and Exige, the Evora, the 2-Eleven, 3-Eleven and 340R. After a personnel reshuffle during the Bahar period, Russell was reinstated as Head of Design in October 2014 and took the title of Design Director in April 2018. Most recently, he has overseen the styling of the Evija and Emira.

During his 35-year career in design, Russell has been involved in numerous Lotus Cars projects, as well as leading design consultancy work for other brands. This rare mix means he has helped to shape everything from forklift trucks to superbikes.

Russell Carr joined Lotus in 1990, and has overseen the styling of the Elise S2, Exige, 2-Eleven, 3-Eleven, 340R and Evora, plus the latest Evija and Emira. (Courtesy Lotus Cars)

Steve Crijns – Evora Exterior Designer

Steve Crijns was recruited by Julian Thomson in 1994 to join the Lotus Design team from the Royal College of Art, where he did an MA in Car Design. Originally from Belgium, Steve worked on project 'Monza' – the Elise Series 2 – and his design was selected for both the exterior and interior. He subsequently designed the Exige S2, launched in 2004. On the Evora project, Steve was responsible for the exterior styling, from the early stages of the programme where package and concept were discussed, via the renderings to clay model phase, through to the first production cars, finalising minute details with the engineers, suppliers and quality inspectors. Steve's work on the Evora created an all-time design classic, and he left Lotus to become Senior Design Manager at McLaren Automotive Ltd.

Steve Crijns and the Evora at the London Motor Show unveiling on July 22nd 2008: his design was selected as the one to productionise. (Courtesy Lotus Cars)

STYLE COUNCIL

Anthony Bushell – Evora Interior Designer

Anthony joined Lotus in September 2001, and his role as Interior Designer at Lotus included the initial design sketches, working closely with the engineering team to make the design feasible, including elements such as materials for colour and trim. Designing the Evora's interior meant creating a cabin that felt sporty, distinctive, yet unmistakably Lotus, though it had more content to package than any previous Lotus.

The Evora cabin architecture was penned by Anthony 'Bush' Bushell, pictured here at the 2008 London Motor Show launch at the ExCel exhibition centre. (Courtesy Lotus Cars)

CHAPTER 3

SEEKING APPROVAL
From acceptance to sign-off

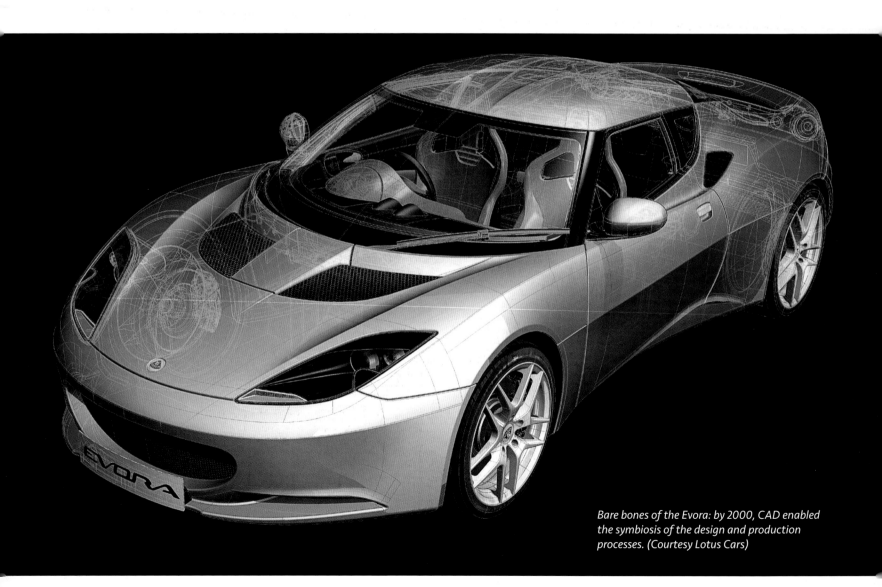

Bare bones of the Evora: by 2000, CAD enabled the symbiosis of the design and production processes. (Courtesy Lotus Cars)

Traditionally, fibreglass clamshells and panels were laid up by hand, but since the advent of the Elise S2 in 2000 they have been made by SOTIRA in France. (Courtesy Lotus Cars)

The IT revolution was well under way in the first decade of the 21st century, and, of course, has advanced in leaps and bounds since then. A decade after the Elise launch, the digital age had matured sufficiently to enable an effective symbiosis between the design and production processes. Where to start? Panels – clamshells – since time immemorial; a stock-in-trade product created largely hands-on within the Lotus factories, were no longer fabricated in-house. The body panel set was commissioned and developed by the French company SOTIRA, a conglomerate based at Mayenne, midway between Rennes and Le Mans, which had been producing all Lotus clamshells and certain other panels using the RTM (resin transfer moulding) process that is basically a variation of the VARI (vacuum assisted resin injection) system Lotus used to use for the Esprit. So, instead of picking a new body vendor and nurturing it through the learning process, or building the body by hand at Hethel as had been the case with the early Elise, Lotus chose a supplier that was a known quantity with an extensive track record, saving time and money and gaining in efficiency.

Even back then, in 2006, Lotus no longer had the facility to make the clams on site. Farming out much of the supply chain threw up more problems, and in May 2007 Lotus was obliged to buy its chassis supplier, Hydro Aluminium, and create Lotus Lightweight Systems in order to protect its supplies. In this instance, Lotus hadn't had to build a new factory to produce the Evora, since it used an existing line: the one originally constructed

Being a new vehicle with different chassis and powertrain, the Evora required a new assembly line to accommodate jigs and two-post lifts. (Courtesy Lotus Cars)

for the VX-220 GM Speedster. The Lotus production line was not quite like that you might find at GM or BMW, where transfer lines and automatic conveyors are continually on the move. At Lotus, the line is a given space in a set of factory buildings, with some jigs and fixtures, and people who know what they are doing. The layout of the Hethel site had to be considered as a whole, for although there were several variants of the Elise in production, their chassis and powertrains were all similar. But, being a new vehicle, larger and with a new V6 engine and gearbox, the Evora required new space, and more plant was built accordingly.

The MRP system – manufacturing resource planning – managed the logistics of Evora production, filtering all the orders from global sales through a single point. There was a built-in time buffer, but if all was running smoothly, three months before a car was needed for delivery, the Manufacturing Department planned its production. They knew how many left- or right-hand drive cars were coming through, how many required air-con and heating, and the various engine specs and body configurations on order. This data was mustered, fed to the Lotus Board for

The headlight was a constant presence in the Evora design – as seen here in the Stanbridge Motorsport GT4 racecar. (Courtesy Stephanie Ewen)

financial approval, and then the Manufacturing Department began allocating the money according to the production schedule. The process was entirely computerised, and the system planned on a rolling monthly basis. Lotus would have preferred a three- to six-month order bank: a year or 18 months was too long for customers to wait, though I recall that Morgan was something like nine years at one point – and any shorter than that put the supply chain under excessive strain.

Once a car went into build, it started in body assembly where it spent about three weeks before moving into the paint shop. Between the two, Lotus allowed a half-day buffer so that it always had the painted set of panels available when the car went onto the line, though that varied if more cars were being built, in which case the buffer time could be reduced. The chassis spent about a week going down the line, being built up in Factories 4 and 5, and then spent another week in the Final

Buy-Off area where fluids and wheels and tyres were added. All the ancillaries, headlights and trusses were waiting in bins to be applied, according to order. Headlamps were configured for left-hand drive, right-hand drive and the American market. About a third of the cars were likely to be right-hand drive, serving the Japanese and UK markets. The lights had to dip in a particular way; while left-hand drive cars had two kinds of dipping methods, left-dip and down-dip, American lights had a dot mark, UK ones had a Department of Transport mark, and European cars an 'E' mark. That highlights the complexity of just one 'everyday' component. In total, there were around 1200 parts per car, but 3500 part-numbers, given the variety of configurations for different global markets. Keeping all that on track was a demanding logistical task.

Some customers began laying down deposits once the Evora was at VP (validation prototype) stage, even though the final retail cost was not yet established. Others awaited news of the retail price and the opportunity to test drive the car. Dealers were champing at the bit, because in the financial climate of late 2008, there were few exciting new cars coming to the showrooms. But the deposits being placed meant that Luke Bennett could plan how to split the left- and right-hand drive orders. The next stage was to start establishing popular paint colours, wheel specifications and other detailing and preferences. This gave him a cushion of about six months' production planning in hand, thanks to keen customers. The production system began loading in late January 2009. The identities of the first 40 cars were split; half were prototypes for the US Federal market programme, which

At certain stages of assembly, the Evora line flanked the Elise and Exige. (Courtesy Lotus Cars)

required compliance with another level of safety legislation. The other half represented a press fleet of 20 cars. After that, consumer cars came on stream, reaching dealers by May 2009.

Customer choice was tightly controlled, such as paint colour, trim, the personalised elements, and so the company invested heavily in the trim shop, headed by Pete Hunter. The development programme was overseen from start to finish by Peter Wainwright, Technical Manager, General Development, who'd been involved with the Evora from its fledgling days. The Test and Development Manager was Peter Lawton, reporting to Chris Dunster. At the beginning of the planning stage, Peter Lawton created a profile of the new vehicle by coordinating its market sector, its performance, its likely public perception, and all the test criteria that the car would need to meet by the time it came to be sold to the public. This profile was then turned into a technical specification against each of the vehicle criteria described in the profile. The technical specification then became the basis for a test programme to meet each of those criteria, in a schedule developed alongside the design and build process.

It was a complicated business: at the beginning of the programme, Lotus operatives had a working list of some 450 tests to complete, and Peter Lawton then had to plan how many vehicles he needed for the test programme, at what level of prototype, and how much budget and resources were required to bring his plan to fruition. He then spent the next two years overseeing the test development programme, ensuring that

Following the Elan Plus 2, Lotus' next four-seater was the 1974 Type 75 Elite, taking the company into a more prestigious market segment. (Courtesy Johnny Tipler)

When locating airbags in the dashboard during crash testing, designers needed to ensure that children could not be harmed by the airbag trapdoor. (Courtesy Lotus Cars)

every compliance and testing requirement was met. With 25 years of experience to draw on, including the Excel, the last two-plus-two built by the company until the Evora, Lawton's next project was the M100, which was supposed to be the new Elan with a Toyota engine, a project canned when Toyota sold its share in Lotus to GM. He then moved on to the anti-lock brake system for the Esprit, before working on the Exige, the Type 121 Europa and the Evora.

Crash test dummies

Pragmatically, the Evora's development was divided up and apportioned to groups within Lotus. These ranged from vehicle dynamics, cooling and thermal management, aerodynamics, crash testing, and NVH – noise, vibration, harshness. During the EP (evaluation prototype) phase Lotus used 23 prototype vehicles, all at different stages of completion. Some were made up of just the chassis, the front and rear sub frames and a few interior parts. And there were approximately the same number of vehicles in the second VP (validation prototype) phase. There were 12 running vehicles, 11 crossover vehicles and nine test tubs, all built to different levels. Part of Peter Lawton's job was to specify each of these prototypes, which components to be fitted, and to which build levels they would be constructed. So, he had to keep in touch with all the modifications that went through the programme, tracking them on a weekly basis, picking up the latest status on each vehicle and identifying where that status needed to be applied. If, for instance, a design changed as a result of one piece of development, his team had to go back and get together two or three engineers to update all the relevant prototypes.

On the rocks

What's the most unforgiving surface you can drive on? I can testify from first-hand experience, it's Belgian Pavé. En route to a pal's wedding in 2019, at Ohain, south of Brussels and within sight of Waterloo's battle site, my satnav directed me down a single-track country lane. Unfortunately, the massive cobbles that constitute Pavé had been churned up by over-large agricultural machinery since Napoleon's day, and my Boxster's suspension was rent asunder: the legacy of this deviation was that all the tracking and wheel alignment were out, plus one bent damper (a Porsche item!) and a scuffed wheel rim. I shouldn't have been surprised that 4x4s were overtaking me via the adjacent fields as I went vehicular bouldering.

So, have no doubt, Pavé is a serious test of a car's suspension and steering, and at least one Evora prototype was subjected to this indignity. The Elise and Type 121 Europa had already endured the ordeal, in which the car drives around 2500km bouncing over these remorseless and unforgiving cobbles, said to represent 100,000 miles of normal road use. It was anticipated that the Evora would be used on a daily basis, and that very high mileage vehicles would be commonplace, so Lotus designed and then validated cars with that in mind. The criteria were that, at the end of the Pavé test, the structure should still hold together, without any major failures. I couldn't say that about my Boxster after just a couple of miles, though to be fair, the cobbles on that Belgian backroad were pretty topsy-turvy! That said, I once drove a Europa S into the Atlas Mountains on a shoot with Jason Parnell, and the car did well to cope with rockfalls and debris that were barely avoidable. Even after what amounted to the equivalent of 10 to 12 year's life in normal driving circumstances it would not be acceptable if the Evora's wishbones dropped off, or the car became unsafe through such a failure.

As for traction monitoring, on-board skid control is mandatory in the Federal market, and the Evora prototypes had to be tested to these standards. Calibrating the car's computer system to identify the behaviour of each wheel against the car's steering angle in order to assess slippage was a huge and relatively experimental task, and Lotus' engineering team took charge of this. Since the Evora was modular in construction, it was relatively straightforward to remove front and rear ends, replace them and use the car for other work. Thus, the vehicle used for the Pavé testing could be re-employed for oil surge and fuel surge work, something not possible in a car with a bonded-on front-end. The Evora's front and rear modules were bolted on, so the technicians were able to make more use of existing prototypes than was possible, for example, with the contemporary Elise where the chassis was a single entity. One of the crashed Evoras was rebuilt into a full development vehicle, and another crashed car was rebuilt to fill the vacancy.

The Evora's brakes had been designed for the stillborn Esprit replacement, and were signed off for the Evora. The anti-lock brake system worked straight out of the box, though there was a development background to that within the programme, wherein the production team created a mule disguised as an Esprit V8, which they dispatched to the Bosch test centre on the frozen lake at Arjeplog in northern Sweden's white-out winter wonderland. I visited the centre during a trip to the Ice Hotel near Kiruna in Lapland, which I report on elsewhere. The brief to Bosch was to provide the concept and initial design work on that vehicle, and the system was transferred straight onto the Evora prototypes.

At the forefront of vehicle dynamics was choice of tyre. The most critical point of contact between any vehicle and the road surface is the tyre, and so Lotus developed a tyre specification in association with Yokohama. Pragmatically, the Evora entered production shod with off-the-shelf Pirelli P-Zeros, though practicality and fine judgement, according to prevailing use, also affect tyre selection – my most recent test car, a GT410, was shod with Michelin Pilot Sport Cups.

Type approval

Every production model needs to gain Type Approval before going on sale. To this end, all production vehicles undergo rigorous crash and safety testing to ensure that they conform

Frequent rockfalls were a peril on the photoshoot when hustling a Europa through Morocco's Atlas Mountain passes. (Courtesy Jason Parnell)

SEEKING APPROVAL

to minimum legal criteria. This means that the vehicle will incorporate all sorts of crash-proof or deformable structures so it's able to contend with internal and external impacts, not to mention compulsory kit like seatbelts and air bags. Naturally, the Evora was no exception. Of the 19 experimental prototypes built between October 2007 and March 2008, six were designated as crash test cars. In the event, only four were needed, because the frontal structure on the Evora can be replaced. So, after a frontal impact test, the team mounted a fresh subframe, new suspension and new front clam, and crashed it again. The same pertained for the rear subframe, designed to be unbolted in the event of such an eventuality, post production.

It's reassuring to know that the vehicle you're travelling in can withstand impacts, which might not be the case in a classic car, and of course to obtain type approval it was necessary to subject Evora prototypes to crash testing. Catapulting the car into barriers to assess deformability of front and back, as well as the integrity of the central passenger cell, achieved injury levels to the anthropomorphic subjects well below those set as the in-house standard. This proved that the Evora was an exceptionally safe vehicle which could be rebuilt after crashes. This work was carried out by Continental Safety Engineering International. Designated prototypes were taped and masked up, with ruler-like stripes over the top and along the sides to evidence degrees of crumpling and deformation. They were then subjected to a destructive crash test programme, even being projected as if from a cannon into a gravel trap. At the other end of the scale, the kerbing test was relatively innocuous. The meat of the crashing involved generating 40 tonnes impact into all sides of the car, simulating 35mph collisions dissipated through the chassis. Some of these aspects were achieved by mounting the car on a revolving gantry, so it could even be assessed upside down.

Monitoring the progress of Project Eagle was John Cooper, Quality Assurance Engineer, responsible for all quality systems within the project. QA covered everything from how the project was run, through to final product quality, and that happened over a number of different stages. As John Cooper put it, "We were actively looking for trouble. In the early days we carried out a process called Failure Mode Effect Analysis (FMEA) where we looked at the design and questioned whether we were going to have problems with the parts fitting, with durability of parts, any issues that the customer would pick up, whether all the in-car systems interface with each other. We prioritised safety systems such as brakes, steering and chassis, and if we identified any potential issues, then we'd captured those before they became an issue. Our work with GM refined that process at Lotus. When we built the VX220 Roadster, GM obliged us to do things in specific ways, and we learned from that, and it became part of our quality function."

Bench mark competitors

Lotus set quality targets by benchmarking its own products against those of its competitors. As a matter of course, Lotus – like all manufacturers – acquires cars from its competitors, and its vehicle auditors go through them to assess the fits and shuts

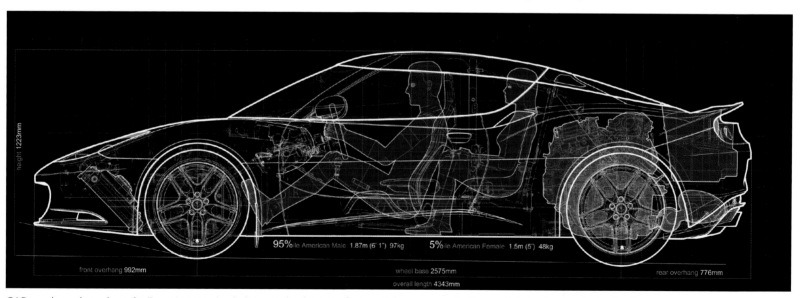

CAD mock-up shows how the Evora is a genuine 2+2, given the driver is 6ft 1in and the passenger in the rear is 5ft. (Courtesy Lotus Cars)

in the exterior, and compare the interior against Lotus' own products, so they know exactly what their quality expectations should be. They then run the new product introduction process with various gateways that identify what points in the programme certain deliverables should be met. The whole time, Quality Assurance was fine-tuning build quality. There's a static assessment, and a water test that ensures they are watertight, then they're taken out onto the Hethel test track for dynamic assessment. "We learned from the current product in order to make the new car the best it could be," said John, "bearing in mind the market sector it was going into. It was a market we had not been in for a long time, and it had certain expectations regarding aspects such as panel fit and closure shut criteria, which the Germanic brands have been achieving for many years. Their cars have steel bodies, while we use composite ones, which presents challenges in itself."

The cabin interior presented issues of its own. "There's lots of leather in this car, and the initial interior concept lagged behind the exterior," said John Cooper, "so we had to work hard to ensure that we hit the standards our customers expect in this segment. The quality of the leather presented some issues at first, and we had to work with our suppliers and let the trim shop know what's expected of them. The manual dexterity required in the trim shop is tremendous, and we identified problems in the base leather, looking at the natural defects that you get in an animal hide."

Double wishbones
The Evora's front suspension consisted of wishbones, as light as the Elise's, but twice as stiff, and very expensive to develop as a consequence. The Evora's wheelbase was longer than the Elise's, but the intention was always that it should provide the same agile Lotus feel and handling. The Elise wheelbase was 90.6in (2300mm) and front track was 56.7in (1440mm), the Evora was 101.4in (2575mm), and front track 61.6in (1565mm), and although the chassis looked huge by comparison, the car's width was controlled by reducing the sill width from 100mm to 80mm, and kinking it at each end of the passenger cell.

Brake setup and handling were honed by Bosch, whose view back in 2006 was that the Evora's brakes were the best in the world. Ride and Handling maestro Matt Becker made the surprising assertion that, "the handling is so good that, even fully loaded, the Evora is quicker around the Nürburgring Nordschleife than the Elise. The objective was to make it handle as well as the Elise, and it exceeded all expectations."

Typically, a team of experts was involved in fine-tuning the Evora's behaviour, whittling away at the specification like

Front corner of an Evora, featuring brake disc and AP Racing calliper, double wishbones and coil-over damper. (Courtesy Lotus Cars)

a sculptor carves and defines a block of stone. They included Ingemar Johansson, Executive Engineer on Chassis Design, Alan Clarke on the brake system, Alex Böss from Bosch, and Dave Tankard, Function Leader for Vehicle Safety. "Which development engineers we had on site depended on where we were in the world," Dave Tankard explained. "We had Alan Jeffers controlling them all, but things like the Pavé test were licensed out, as no one really wanted to do that as it shakes your fillings out, shuddering along the cobbles hour after hour."

A few years ago, *Top Gear* featured the half-mile stretch of the Belgian cobbles at Chobham test track – Longcross to you film buffs – and all sorts of things were falling off. "We drove 2200km of Pavé in the cars," recalled Dave, "which is the same as GM would do on the next Astra or some other everyday car. We're used to driving the cars for engineering clients, so we know what's required. Things like the corrosion test, which is the wet weather equivalent of leaving something for ten years in Death Valley, but made wet by dropping it into the sea, leaving it there and coming back a few years later. Photos don't do it justice; they don't look the same in 2D as they do in real life – corrosion can do horrific things!" As Function Leader for Vehicle Safety, Dave Tankard's commitment to the Evora project ran through the programme from conception to production. Working

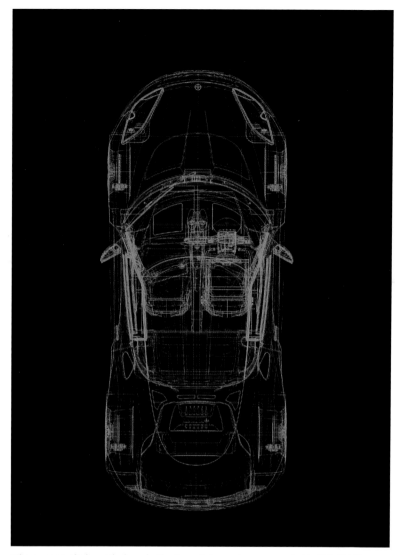

Like a very orderly spider's web, the Evora's innards are revealed in this CAD image. (Courtesy Lotus Cars)

Pre-production CAD image hints at the Evora's underpinnings. (Courtesy Lotus Cars)

closely with Dave Marler, function leader for CAE (computer aided engineering) analysis and his team, Dave Tankard was responsible for the restraint system, which included the seatbelts and their interface with the seats and vehicle structure, driver and passenger airbags, and he also managed the crash test programme.

Safety aspects regarding the car's architecture were addressed in the course of the earliest planning meetings, such was the need for a collapsible steering column, maximising the internal space in the cabin to prevent injuries in the event of an impact. Thus, the Evora has a Federal-compliant steering column that's supported over its complete collapse length. The ECU airbag controller and crash sensors were also packaged within the car. Dave Tankard and Brian Hope from the Homologation Department implemented the relevant safety requirements into the programme, describing exactly where airbags would be mounted, and the kind of seatbelts that would be required. Dave also contributed to the seat specification and its performance criteria, noting the level of clearance required under the front clamshell to reduce injury to pedestrians in the event they were knocked over and bounced onto the car, and showed where bumpers needed to be positioned in order to meet pedestrians' lower leg safety requirements.

Before the styling was signed off, the Vehicle Safety team was marking out the pedestrian impact zones on the models. The models were then scanned, and the data thus derived incorporated into the CAD systems. When considering the fitment of airbags into the dashboard panel, Dave stipulated the area on the front leading edge of the instrument panel necessary to make sure that children were not in front of the airbag trapdoor. In deciding on a steering wheel design, the supplier advised Dave of the package requirements for the driver's airbag and inflator, so that could be incorporated into the design of the steering wheel.

For the first time, Lotus developed a car to obtain type approval for the whole vehicle from the outset, to meet European and US Federal standards, bearing in mind the Evora was always destined to be a global model. The Evora was originally equipped with Autoliv seatbelts, sourced from Autoliv Hirotako Malaysia via Proton. The driver and passenger airbags were developed by Lotus in a joint venture with Continental Safety Engineering International, with help of Project Manager David Haywood and Martin Böhme. Cars complying with European standards had to meet a 35mph offset deformable barrier test, though unofficially, NCAP (European New Car Assessment Programme) testing took place at 40mph. "The legal test is conducted at 35mph," said Dave Tankard. "US Federal requirements have risen from 30mph up to the European legal minimum of 35mph, but into a flat, rigid barrier, so the energy has increased by nearly 40 per cent, and since more impact energy has to absorbed for the US spec cars, that is what the Evora's front-end is designed to accommodate." Before the actual crash testing exercise began, Lotus engineers mounted the crash structure, which basically amounted to the front and rear of the car on a four-wheel buck, known as the trolley, ballasted to equal the mass of the vehicle. In an impact of alarming ferocity, this was propelled into the unyielding wall to replicate the effects of a collision, and the data gleaned gave the team information as to the likely outcome of the actual crash tests.

The Evora was designed so that the central tub of the vehicle was a safety cell, where there was little or no deformation, with all the deformation within the subframes, both front and rear. Lotus tested the crash structure even before any tubs were made. "The structures incorporated the bumper beam and subframe, with fabrication representing the stiffness of the front suspension as it would short-out the crash structure, all hanging off the front of the trolley," explained Dave. "Then the trolley was impacted at 35mph into a rigid wall, and all the energy was absorbed by the crash structure. Computer simulation could then correlate the model to the test, and there was also a visual advantage because there was no car body in the way to obscure the effects. The crash structure was analysed frame by frame, and it revealed a progressive crush, working its way back to the equivalent of the front of the tub."

During the crash test programme some cars were tested three times over, with two vehicles rebuilt within two days to make them fit for purpose, though not for on-the-road driving

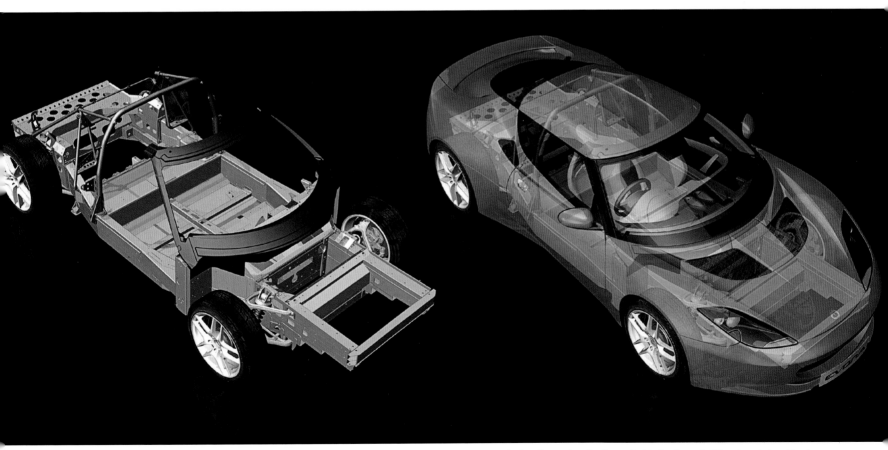

CAD renderings of the Evora rolling chassis, showing roll-over superstructure and screen surround, plus the entire platform clad in bodywork. (Courtesy Lotus Cars)

Evora steering wheels were designed and made in-house; in this case, the IPS version with paddle shifts. (Courtesy Lotus Cars)

capability. The crash facility provided real-time video monitoring, which could be housed under a glass plate in the floor, mounted on an overhead gantry or to the side of the vehicle, filming the event at 1000 frames per second.

The first batch of EP experimental prototypes was subjected to 12 crash tests conducted in the first phase of the programme, including a side-impact and an 80kph (50mph) rear impact to verify the integrity of the fuel system. In the second phase, again at 35mph frontal impact and 50mph rear impact, there was no significant deformation in the passenger cell, nor to the suspension systems either. Apart from the absence of trim, the EP prototypes were similar to the ultimate production model, so the findings were absolutely relevant. The crash-test prototypes endured the treatment surprisingly well. Parts were recycled and new crash structures installed so they could then undergo the high-speed impact test, and then the Federal fuel system intensity test. This was where the vehicle was tested for any fluid leaks immediately after the high-speed test, so the mule was effectively skewered on a spit and rotated through 360° and held there for five minutes at 90° intervals to check for any leaks. There were none at all.

The European crash test car was projected at 35mph into a deformable barrier, in such a way that all the load was transferred to one side of the vehicle. This caused the car to cant over in one direction, and although the upper and lower wishbones showed some evidence of deformation, this damage was not passed onto the body tub, and both doors opened normally after the test. The result of the test was close to the prediction from the computer simulation. The European-standard side impact test revealed how much energy the Evora could absorb or, literally, how much crush it could take. It was a careful engineering balance between a really stiff vehicle, which ended up with very high deceleration, and a more progressive energy absorption, where the occupant stayed comparatively still. Another milestone in passenger security was seat design, which was also a crucial safety factor. During tests, the Evora seat was bolted to the side of the mule that took the hits during the European side-impact tests, and its relatively flat design spread the energy load. By the same token, the car's strong chassis side structure absorbed much of the energy punched back into it in the milliseconds after impact. The 50mph rear impact test was a very severe crash, but it didn't measure potential injury levels. The back of the engine didn't even move, and all the potentially devastating energy was absorbed through the rear subframe. The purpose of this test was to establish fuel cell integrity. The fuel tank was mounted in the middle of the vehicle, and there was no leakage at all as a result of the crash test, even though legal requirements allowed for some leakage.

Next came the abuse testing. The car was run over Belgian Pavé and other unlikely hazards identified as block track bumps, washboards, railway crossings and country jumps, straddle paths and potholes. This was all done, along with the crash testing, at Continental Tyre's facility in Germany. The abuse test results were correlated with the results of the low speed impact tests where the airbag was not fired, and the outcome was an algorithm which ran the crash sensor so that the airbags would not fire under certain conditions, but would under others, meaning that it would only fire when required. After a rally-style ramp jump the vehicle remained driveable, despite the severe shock it had just suffered. In the first phase of testing the airbags were fired remotely to guarantee that both they and the seatbelts would perform within a certain timespan, say 20 milliseconds. The airbags were correctly calibrated for the next phase so that the tests would then fire against the crash pulse in the right scenario.

From 2005, all new vehicles had to meet new requirements for pedestrian safety. So, Dave Tankard and his team marked out an appropriate zone on the car's bonnet, extending 1 metre behind its leading edge and around the bottom of the windscreen, and the car had then to be impacted at various locations. Assuming that a pedestrian was struck by the front of the car, a child's head-form was launched at the bonnet, and the desired result was a zone where a third of the impact was below a hit of 2000 head injury criteria, and the remaining third of the impact a hit below 1000, which was, in practice, a dent of between 6 to 9 inches in the bonnet. The advantage with the Evora was that since there was no engine under the bonnet, and the bodywork was a composite material and therefore highly deformable, the impact caused little damage to the car since it was able to spring back into shape, while at the same time providing an energy-absorbing pad for the impact dummy.

The Evora did not have to conform to a further pedestrian test, known as 'the upper leg,' because it was a sports car, but it was required to conform in an area related to 'the lower leg,' another zone that was marked out accordingly on the front of the car. The Evora also had to meet lower leg acceleration and bending requirements, where the legal limit was 200g and 15-degree angles. To get good results for the lower leg impact tests, standards actually improved, insofar as fewer injuries were caused without resorting to the use of protective foam, by allowing the pedestrian's leg to deform the bumper skin. A lower lip on the front bumper caught the leg. Sounds gruesome, but the ankle impact area was supported and energy absorbed, so that the unfortunate walker's leg didn't bend in half.

Crucially, the Evora's steering wheel was designed in-house. It incorporated a magnesium armature, which could be changed to make the diameter smaller so it could also be fitted to an Elise. It featured optional cruise control buttons, or blanks depending on the customer's selection of facilities, in an aluminium finish. In the event of an impact, the flap would open to release the detonating bag. The passenger side airbag was tucked behind a steel door with a mechanical hinge with a foam cover. The advantage of this steel-door system was that whatever surface the design team applied to other parts of the car interior, airbag performance would be unaffected. This was an unexpectedly important result, since the variation of hide quality and stitching, plus the potential to degrade in strong sunlight, made a leather upholstered airbag lid a highly unpredictable prospect for airbag detonation. Steel airbag doors were preferable to aluminium in this location because, in static deployment tests, steel reduced the velocity of the door opening. By tuning the bend in the steel lid and putting in different slots, the stiffness of the hinge could be controlled, and thus the velocity rate of the airbag door opening. In one crash test car I saw during a factory visit, the occupant was a small female dummy, as required by Federal standards. All injury levels round the neck could be measured, and whiplash injuries after a head-on impact test were considered acceptable. Lotus was not obliged to contemplate issues to do with head restraints in such circumstances, since these were dealt with by seat makers Recaro.

Luke Bennett – Manufacturing Director

From Production Operations Manager, Luke Bennett became Operations Director for Lotus Cars Ltd in 2007, controlling all the manufacturing processes used in the construction of Lotus models and third-party cars assembled at Hethel. He was also a Board Director of Group Lotus PLC. In the case of the Evora, he led the most advanced assembly process ever undertaken by Lotus, one of the most sophisticated for low-volume niche car production, including the introduction of Visual Process Control systems, ensuring that the car was built to the highest quality standards.

Luke did a four-year apprenticeship as a toolmaker in the plastics industry, followed by two years in product design and a year in quality management before joining Lotus' Manufacturing Engineering in 1990, where he was active in the launch of the M100 Elan. In a drive to improve efficiency, he played a key role in leading Kaizen workshops – literally, a quick but structured method of improvement that can be implemented with minimal capital investment – saving millions of pounds and kick-starting Lotus' Continuous Improvement experience. He was closely involved in several more product launches including Elise S1 and Esprit V8, helping define company strategy leading to the launch of the Elise S2. He was also instrumental in raising Hethel's annual output to 7000 units, including the GM Speedster/VX220. Lessons learned from the introduction of the Speedster/VX220 were translated into the US launch of the Elise, paving the way for the Evora. Luke is currently Executive Director of Manufacturing at McLaren Automotive.

The GM VX-220 Roadster, pictured here at Abbeville circuit in France, was another vehicle based on the Elise platform and built at Hethel. (Courtesy Johnny Tipler)

Fully assembled cars, complete with cabin interiors and fitted with wheels and tyres, heading for the rolling road. (Courtesy Lotus Cars)

The final airbag tests were especially relevant, bearing in mind the Evora's 2+2 cabin. These child-oriented trials were the 'three-year-old' and 'six-year-old' tests. Again, emotionally disconcerting, but absolutely necessary. The concept was that the airbag pushed the child backwards to avoid harm or injury, even when unrestrained by a seatbelt and standing in front of the passenger airbag. Both ages were mandatory for the US market. Equally stringent testing was applied to the seatbelts. Safety issues are not sexy, and they can't be overlooked nor taken for granted, so it's reassuring to know that so much work went into making the Evora shell as safe an environment as possible.

The EP and VP prototypes

A prototype vehicle is effectively a tool of the product design process, which allows the designers, engineers and other members of the project team to test their theories and designs. This enables the vehicle's performance in all areas to be confirmed before committing the new project to production. A new model such as the Evora starts off with a product brief, which evolves into a concept description, a product profile, and a detailed vehicle technical specification. These become CAD (computer aided design) drawings, initially of the vehicle package, which defines the basic layout, number of occupants, powertrain location, fuel tank and luggage space, to generate the footprint of the vehicle. This package provides the data to generate styling renderings, scale models and, finally, the full-size clay model. While the style of exterior and interior is evolving, the engineering team develop the structure, suspension, body exterior and interior construction, and detail vehicle systems layouts. During this period, mules are built using donor vehicles and parts to evaluate some of the mechanical aspects.

The first prototype phase cars, designated EP – evaluation prototypes – are close to production, but manufactured to facilitate the creation of prototype tooling. Body panels are a good example: the EP bodies for Evora were laid-by-hand fibreglass composite, fashioned from tooling, machined from Ureol, a composite polyurethane modelling material, compared to the later VP – validation prototypes – whose panels were manufactured from injection-moulded fibreglass composite, derived from the chromium-plated steel production tools.

For the EP phase, Lotus built 19 complete vehicles and several bucks. These were used for the development programme, which included crash testing, structural durability, ride and handling, powertrain and vehicle systems development. This first phase of development provided feedback to the design engineers for the release of production designs. One EP prototype build was developed to a very advanced stage so that it could be used as the first show car. The second prototype phase, designated VPs, are normally built from final production design parts using production tooling, jigs and fixtures. This allows the production assembly process to be proven, and then the vehicles built by this process repeat the EP development programme, validating the production parts and vehicles. For the Evora, this phase employed 23 vehicles for development and crash testing, plus bucks for testing individual components, such as door assembly, window lifts, locking and latching systems. The Evora was destined to be a World Car from the outset, so the prototype programme included vehicles destined for field testing in those territories where climates were extreme. In addition to an extensive European-based programme, further testing of EF – Federal prototypes – was undertaken in the USA, visiting many States including Colorado, Texas, Nevada, and California. Hot and cold climate trips to Scandinavia and Australia also formed part of the acclimatisation programme.

Approaching the finish line: Evoras are still on trolleys, but contain powertrain and suspension. (Courtesy Lotus Cars)

CHAPTER 4

UNDERPINNINGS
Keystones of the platform

When the Evora was conceived, Lotus already had a decade's worth of experience making cars on extruded aluminium chassis, from the Elise and Exige to the Europa, VX220 and Tesla Roadster. Logically, the same technology was broadly applied to the Evora. Whenever practical, Lotus utilised its existing suppliers in the design and manufacture of the Evora, and the chassis producer was fundamental to the plan. Until relocating to Norwich in 2020, the chassis was produced by Lotus Lightweight Structures' 120-strong Worcester-based workforce, having become part of Group Lotus Plc in May 2008. Formerly based at Tonder in Denmark as Hydro Raufoss Automotive, specialising in aluminium extrusions for window frames and greenhouses, the firm set up an automotive division when it came on board with Lotus. After the Elise was launched in 1996, the business came to the UK in 1998 and traded as Hydro Aluminium, making the chassis for the Elise and its siblings. Lotus bought the company in order to secure its exclusive aluminium extrusion technology and chassis supply chain. The entire chassis-manufacturing operation left

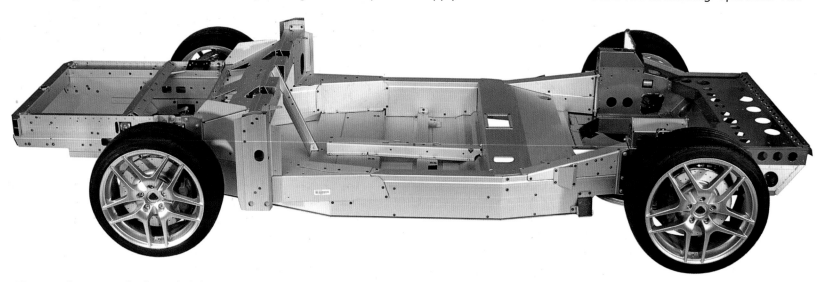

The Evora chassis is made of extruded aluminium sections, bonded and riveted together, comprising a central module, with separately crafted modules attached to front and rear. (Courtesy Lotus Cars)

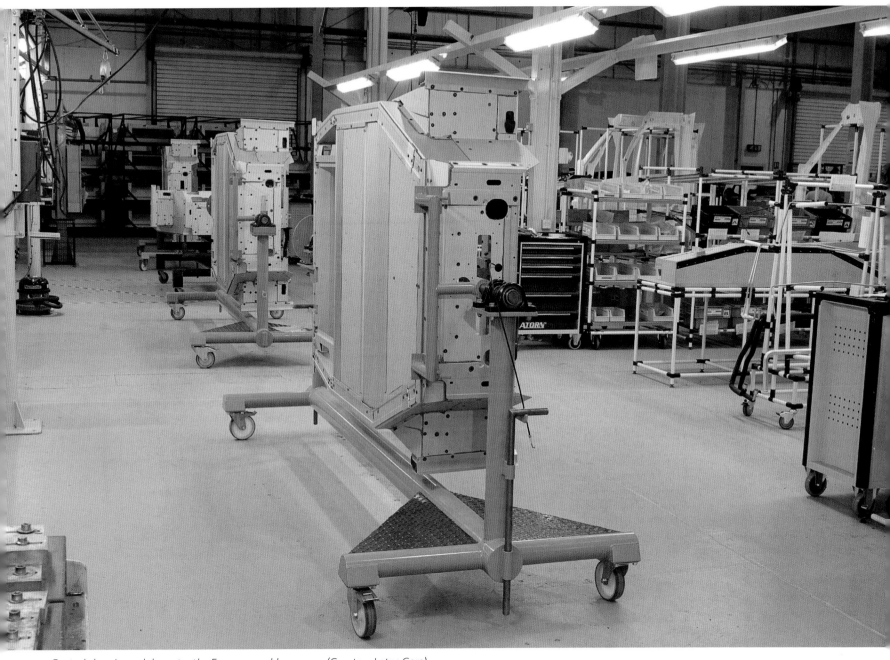

Central chassis modules enter the Evora assembly process. (Courtesy Lotus Cars)

Worcester and relocated onto the trading estate near Norwich Airport in 2020.

Aluminium extrusions are lengths of aluminium passed through a former, rather like how a tube of toothpaste works. Other types of chassis construction include the multi-tubular spaceframe employed by Lotus racing cars such as the Type 16 from 1958 and the ongoing Seven sports car: complex, lightweight, and with great torsional rigidity, but a specialist product. The monocoque: largely sheets of welded aluminium forming a tub, like the Type 25 F1 car, or in carbon-fibre like the Type 87 F1 car; and modern saloon cars which are formed of a welded shell of steel panels. The Evora's chassis was made in the same way as the Elise, of extruded aluminium sections joined together by adhesive and rivets. When launched, the Elise chassis was revolutionary in the automotive industry, using bonded aluminium extrusions and rivet reinforcements for the

The Elise chassis is one entire unit, composed of aluminium extrusions, bonded and riveted together. (Courtesy Lotus Cars)

first time. In partnership with Ciba Polymers of Switzerland and Lotus, Hydro Aluminium discovered that the component parts could be bonded by epoxy resin, provided the bonding site was correctly pre-treated. Rivets ensured the joints were bonded accurately. A significant benefit was that the extruded sections could be half the thickness (1.5mm) of chassis members joined by traditional aluminium welding techniques, because welded joints are inherently weak and demand thicker members so joints will be strong enough.

Godfather of the extruded aluminium chassis, designer Richard Rackham explained the science: "Bonding joints has many advantages over welding. It's more precise, because it eliminates the distortion that comes with welding. This is crucial in a high-performance car structure because the point where the suspension is joined to the chassis has to be controlled to within 0.5mm, or there will be a great variation in handling between vehicles. Also, the heat generated in the welding process has an adverse effect on the aluminium. Using the bonding method enabled Lotus to take advantage of the strength-to-weight benefits provided by heat-sensitive aluminium alloys, which could not be welded easily without degrading some of their properties. Bonding also spreads the loads across a greater area than welding, giving greater strength." Overall, it makes low-volume production more economical, and provides the highest rigidity-to-weight ratio available apart from a carbon-fibre monocoque. As Richard pointed out, "The tooling to achieve complex-shaped extrusions costs only a few thousand pounds whereas tooling to press aluminium sheet costs hundreds of thousands of pounds." Moreover, an equivalent chassis in steel would weigh twice as much.

Richard Rackham admitted that his basic design concept for the Elise chassis mimicked Lego to a certain extent. The extrusions were designed to accommodate the bonding process, each linked with adjoining members by a tongue-and-groove overlapping joint, the securing rivets making a Lego-like fastening. Broad, flat areas make provision for bonding sites. Extruded members feature 0.2mm ridges along all adjoining surfaces to ensure that the adhesive is not squeezed out during the bonding process and maintain the gap width at 0.5mm. The Evora's main chassis module consisted of 35 extruded components, formed of 6063 aluminium-magnesium-silicon alloy. The bonding adhesive is a single-part, heat-cured epoxy paste (XD 4600). For the record, its tensile strength is rated at 60MPa and it has an E-modulus of 3500MPa. It has a paste-like consistency until set and is very stable, and the curing process takes 40 minutes at 200°C.

Richard elucidated on the chassis concept in simplistic terms: "The Evora chassis is essentially nothing more than a complex bracket that accommodates the vehicle systems in a compact order, such that the exterior surface can be stunning and the vehicle dimensions minimal; of course, whilst meeting the stiffness and legislative requirements." It gets more complicated. "The Elise was a pioneering exercise in bonded aluminium construction, so the Evora has been able to build on that expertise, and on developments over the last 12 or 13 years to build a chassis that is more efficient and 2.5 times stiffer. The Elise was stiff at just under 10,000Nm, about 9,8000Nm per degree of twist, while the Evora tub is rated at 26,000Nm per degree. Stiffness has been achieved through the manufacturing process, but it is also a preconceived by-product of other componentry and of exploiting opportunities for stiffness wherever they present themselves."

That said, the design was largely driven by repairability and safety, but because of all the carefully managed load paths around the occupants in the car, stiffness was enhanced. Work on joints proved so successful that Lotus took out patents on its

Access to the cabin, front and rear, meant ergonomics were crucial in determining the precise location of seats and controls. (Courtesy Lotus Cars)

design for the joints between the front subframe and the main tub. The load paths functioned well, and the joint conditions were good, so together they combined to give an extremely efficient vehicle structure. Additional load paths ran up through the windscreen, through the roof panel and back through the seat anchorage frame, increasing structural performance.

These load paths were revealed by computer analysis, looking at stress, strain energy and different torsional situations. The team looked at the extreme conditions the prototypes underwent during testing to show where the high loads were going into the car, thus ensuring that the materials and the load paths were more than adequate. The analysis showed where there was any material redundancy that could be removed, or if it needed to be increased in other places. This diligence gave the structure maximum efficiency, since every element is performing in its own way to share the loads.

The Evora had a new modular chassis design, incorporating subframes front and rear, whereas the Elise only had a rear subframe, so the Elise chassis was effectively made up of two pieces, while the Evora was three. Consequently, the Evora chassis was more complex and heavier than the Elise design, though of course the Evora has a relatively larger cabin. The cabin-forward design presented a weight distribution of 60 per cent in the rear and 40 per cent in the front, with the Evora's wheelbase 275mm longer than that of the Elise.

While similarities and comparisons with the Elise chassis were inevitable, the Evora started with a clean sheet of paper. At the outset, a benchmarking exercise was carried out against a range of comparable vehicles, including the BMW Z4, various generations of the Porsche 911, Cayman and Boxster, the Audi TT and R8 and Ferrari 430, as well as the Elise, of course. Multiple measurements were taken, but Lotus also held three consumer scoring exercises in the Styling Studio, where 150 men and women of different ages and measurements tried out a range of cars and scored them for usability. This information was translated into a Project Eagle cockpit buck, made of wood and foam, and comprising the seats, steering wheel and roofline, which these same 150 people were invited to try again. As a result, Lotus generated the dimensions for the chassis, incorporating the crucial ease of ingress and egress to and from the car. This scoring exercise was repeated during the design process as elements were altered to try and achieve a higher score. Lotus clearly wanted to be at least as good as its competitors, and preferably better on factors such as cabin access and internal ergonomics, including providing the best possible driving position.

The seating buck was an interesting object in its own right. The inside was a correct representation of the cabin interior, while the outside was confined to the necessary elements. So, it had the roof panel, which could be detached, the doors opened and closed, it had pedals and a steering column, console and dashboard. Fine-tuned and worked up on CAD, it had standard mannequins packaged into the car from day one, measuring 6ft 5in up front, on the basis of CEO Mike Kimberley's tallness.

Richard Rackham was the Lotus Vehicle Architect, a title unique in the car industry, and a role he continues to perform, having overseen projects including the Tesla Roadster, the Vauxhall VX220, Aston Martin's VH platform cars, McLaren, the Evija and the Emira. In high-volume manufacturers, Richard's task would be shared between more individuals, but at Lotus, people tend to do a lot more within their job spec and have a broader awareness of what's going on in the rest of the company than might be the case elsewhere in the industry. Different pieces of the puzzle were created and brought from across the company to create the new car. Richard's team began configuring the chassis as the styling concept was developed and the marketing people were working on the product profile, markets and customer expectations.

Once the need for the new model had been identified at Board level and its basic profile established, Richard began to calculate how much the vehicle might weigh, a figure that would give an idea of the appropriate tyre size and width. On full lock the arc defined by the front tyre generated a dimension from which the footwells within the cabin could be configured, all of which contributed to the definition of the front-end. Gradually all the building blocks of the package emerged. Factors such as the occupants, the powertrain and ancillaries such as fuel tank and cooling system, the wheel envelopes – defined by how much space the wheels took up as they go up and down. Such fundamentals had to be coordinated in a way that met the product profile, and a package of information emerged that gave the styling team a basis to work on and wrap some realistic lines around. The information flow between Vehicle Architecture and the Design studio was constant. For instance, Design might well have told Richard Rackham that the windscreen needed to be in a particular location, and Richard would then have researched which part of the glass legally needed to be wiped, and depending on the occupants' field of vision, specific areas of the screen needed to be cleared by the wiper blade. He was then able to ensure that the design of the wiper mechanism was legally compliant and feed that information back into the main package.

The size of the radiator was another example; the engine configuration dictated the dimensions of the radiator core,

Releasing 2000bhp, Gavan Kershaw lights up the rear tyres of the Evija hypercar on the Goodwood startline, 2021. (Courtesy Antony Fraser)

which in turn dictated the size of the holes in the bodywork, a significant consideration for the styling team. Richard had to make sure that it worked harmoniously with safety factors such as pedestrian impact legislation.

As we saw in the previous chapter, the Evora frontal structure and bodywork was designed to be relatively soft, with an ankle catcher at the bottom of the bumper which ensured the pedestrian would fall onto a soft bonnet. That softness was also a function of the radiator ducts, which emerged in the upper deck of the panel, while the wiper mechanism had to be accommodated aft of them. There was much more energy to disperse on the Evora than on the Elise or the Exige because it weighed half as much again as either of those models, so the tub was more solid and significantly stiffer. The front and rear structures were essentially sacrificial, as shown on the crash testing video of a perpendicular frontal impact, where the subframe concertinaed to about a third of its original length at 35mph. This came about by the spacing of the rivets, the thickness, caging and shapes of the boxes, and the positioning of the windows. Richard credited Tony Shute with locating the battery package over the gearbox and making it accessible from the boot, "A great solution, since it is in a dry but comparatively cool area of the car."

The Evora's power unit is the Toyota-sourced 3456cc VVT-i dohc V6, fitted with Lotus' engine management system, and producing 276bhp in normally-aspirated format. (Courtesy Lotus Cars)

During the planning phase, before the engine was selected, the team worked with generic transverse V6 information from several manufacturers. The process of sourcing the engine meant Lotus could investigate a range of products; ideally ones in the next generation of development. Lotus needed to be able to adapt the engine to meet its own needs, something few major manufacturers permit but, happily, the relationship with existing engine supplier Toyota allowed this option exclusively to Lotus, and soon enough Lotus was able to make a selection. Meanwhile, the team had to get on with packaging the car and laying out the architecture, and the engineers had to engineer the structures around an engine. To begin with, they used a generic V6, which was less than ideal, as the precise shape and dimensions of the eventual exhaust system were not known; nor exactly where the pipes would exit the rear of the car.

Richard Rackham had to constantly resist the temptation to make the car bigger, simply to fit in more. He commented on the monstrous size of modern supercars like the Pagani Zonda, which has space to spare within the chassis, making it an easy option as an engineering exercise, rather than a smaller, more intensely engineered and specially-efficient vehicle like the Lotus. The Evora's wheelbase was only 11in (275mm) bigger than the Elise's, but it had three inches more seat travel – the source of the additional space for rear seat occupants – not to mention an extra bank of cylinders, assuming a V6-configuration rather than a transverse-mounted in-line four. The implication was that everything else had to be squeezed up. While there was a generous air gap around the fuel tank on the Elise, on the Evora that gap was reduced to less than 20mm in the transverse location under the rear seats.

By definition, a Lotus had to be nimble, so all the elements had to be pulled tightly together – a demanding intellectual feat that Richard Rackham managed to accomplish.

"The construction risk is increased when the design has been tightened up in this way," he observed. "All the engineering tolerances have to be more precise, because the engine reduced clearance around it, and the amount of air that can pass through the vehicle has been reduced." This has been a difficult challenge for the engineering team in particular, demanding elevated design and package management.

Heat management was not a new problem for Lotus, partly due to the fact that it had tended to close off the underbodies for optimum airflow, which generates a certain amount of aerodynamic efficiency, so the underneath of the car was flat, with some strategically-placed holes in the underbody to allow air in. That was how the Elise was configured. The vehicle structure on the Evora had to be kept below 100ºC, because that was the critical temperature needed to sustain the longevity of the bonded elements. The Evora carried three catalytic converters, and their temperature ran to 1000ºC, which the bonding adhesives and rivets on the aluminium structures were not supportive of, so the catalysts were contained within the galvanised steel rear subframe and shielded in areas with chimney vents.

The Elise was a racer for the road – or, at least, its sibling, the Exige, was – entailing a certain amount of contortion to enter and leave the cockpit, but the Evora was intended to be

UNDERPINNINGS

Early on in the assembly process, the wiring loom and pipework are installed, while the front sub-section has the radiator mounted in the underside. (Courtesy Lotus Cars)

a Lotus for everyday use, so one primary concern was the ease of getting in and out of it. Although the chassis concept was the same as that of the Elise, the Evora's structural side members were kinked, and extended further out than the Elise's, and a torsional member running along the back and around the fuel tank is smaller to accommodate the rear occupants of the 2+2. So, both elements were modified to suit the Evora's product profile, and although they resulted in a weight gain, the car's everyday usage potential was infinitely improved.

Another factor in the usability spectrum was ease of repair. With the Evora's separate front and galvanised steel rear subframes, replacement was no problem, and that proved its worth during the development programme when test cars were crashed, rebuilt and reused in quick succession. Another reason why the Evora was more modular in construction than the Elise was because that helped Lotus keep its options open if it wished to build a bigger, heavier car. That way, the front and rear modules that carried the driveline and suspension would be much easier to upgrade.

The balance of a mid-engined car required that the weight be kept as centralised as possible. "Although the fuel tank is to

Racer for the road: the Exige S that I drove to Portofino, exploiting its pin-sharp handling along the serpentine roads flanking the Italian Riviera. (Courtesy Jason Parnell)

The main section of the Evora chassis enters the assembly process, rotated from vertical to flat to enable construction of the car. (Courtesy Lotus Cars)

the rear, there is no weight hanging around the rear wheels," observed Richard, "since the tank, as well as the engine and gearbox, are all in front of them. There is also barely anything ahead of the front wheels, so most of the weight is in the middle of the car, the distribution being about 40/60."

It took Richard Rackham and his Concept Engineering Team around six months to clear the concept design for the chassis and the front and rear subframes. Then the Release Engineer, Ingemar Johansson, and Design Engineers, Chris Dunster and Adrian Gerber, developed the production designs and the designs for the test series in consort with Lotus Lightweight Structures and Manufacturing Engineering. As soon as the concept designs for the chassis structures were selected, the Design Engineering Group worked with the Manufacturing Engineering Group and Lotus Lightweight Structures to establish the production solutions for the chassis tub and the front and rear subframes.

When progressing from the initial chassis design to the final product, all the sizing was done by computers. The process of evaluating crash tests and pedestrian impact tests was typical.

The tests were first simulated with computer-aided engineering – CAE – and the test itself proved that the CAE and its assumptions were correctly evaluated by the computer. The cost-benefit of reaching this stage of accuracy was already appreciable, since Lotus was able to reduce the evaluation testing and to crash fewer cars to prove that its analysis was correct before going for validation testing, after which came a Certification test.

The importance of CAE right through the Evora build programme was applauded by Dave Marler, Executive Engineer, Vehicle CAE. He described how the Evora's crash durability and structural stiffness was achieved: "Within the CAE function, we look after the structural integrity of any part of the vehicle, so suspension and chassis, and the body from the point of view crash stiffness, durability of all aspects of the structure. I try and promote proactive CAE, which means we use the analysis to help drive a design, not just follow a design, so we integrate the analysis into the project from the concept stage." All these parameters needed to be tied up before they could even begin building an EP car. "This work was happening right at the start of the

The Evora's A-pillar, body sides and roof panel all contribute significantly to the car's structure and torsional stiffness. (Courtesy Lotus Cars)

programme," said Dave, "where we were developing the models and driving the design, and the whole purpose of the analysis was to minimise risk when you actually start making vehicles, so we tried to get the stiffness and the crash characteristics working as best we possibly could on the computer, so that when we got into a physical test we were confident that it was going to achieve the results that we predicted."

The basic elements of the vehicle's structure were established in the first six months, with CAE conducting its analyses continuously until EP release. "We were updating the models and making small design changes but the fundamental sizes of the extrusions for the crash structure in the main tub were released very early on, probably four or five months into the programme. You define the basic die sections, and then you can finalise the machining operations, so it is like a three-phase release to define the key parameters of the structure."

The extruded aluminium chassis was pioneered in the Elise S1 of the mid-1990s, and since then, Lotus has employed aluminium extrusions for its vehicle chassis. Apart from its efficiency, this is also because typical tooling costs, even for some of the big extrusion dies, is in the order of £1000s, rather than £10,000s or even £100,000s, which large steel presses used by major manufacturers cost. "Extrusions are produced in continuous lengths," Dave Marler explained, "and then they are chopped off and machined to obtain the desired characteristics. Then there are things we can do in terms of sheet-folded reinforcements and toe-box areas. There are over 30 extrusions in the vehicle, and all those characteristics were finalised fairly early on, so that even after we'd gone right through EP and VP, we had probably only changed two extrusions, and those were for the provision of additional features and not necessarily structural. So, we got pretty good predictions up front, and the structure didn't have to change greatly from that.

"In CAE, one of the first things we were interested in was the length of the front crash rail, the offset between the crash rail and the sill member, so we could determine the basic structural requirements of the package of the vehicle. Obviously, the crash rail is set inboard as a result of the tyre-wheel envelope, and that is what determines these characteristics. On the Evora, we needed to increase the package space to have bigger seats and more foot room, and that inevitably meant a kink in the chassis. One of the things we had to establish was how much crash space we'd got in front of the toe box area, what was the offset between the crash rail and the sill, and also the height between the crash rail and the sill, because an impact generates a bending moment. You have then got to balance the characteristics of the structure to minimise the acceleration loads on the occupants, and how much crush space you have got in the vehicle, and this had a major impact on the styling of the vehicle, because we would say, 'We need a front overhang of so many millimetres,' and the stylist would say, 'I want 100mm less,' so you then have a trade-off between styling and vehicle architecture, and you come up with a compromise between the restraint system and the capability of the crash structure, and how that impacts the styling. We needed about 600- or 700mm of crush in the front rail to achieve the characteristics that we wanted."

Dave Marler's passion for Lotus began with a Series 1 Seven when aged 15. In addition to the crash performance, his CAE brief included Evora's stiffness characteristics. "In order to have traditional Lotus levels of drivability you need a very stiff structure," he affirmed. "That includes stiff hard points for the suspension, so another major part of our work was to look at local stiffnesses. The bending stiffness is very heavily dominated by the sill depth. By looking at how the structure behaves, which parts of the structure are working hard, and using various optimisation techniques, we can define the size of these sill members to give the stiffness characteristics that we want. The toe box area and the tank bay area, and around the rear of the subframe, all contribute greatly to the torsional stiffness of the vehicle. Our target for Evora was to match the same structural feel of the Elise, so our targets for stiffness and model characteristics were considerably greater than the Elise, purely because it has a longer wheelbase and is a bigger, wider car. With the Evora structure, unlike the Elise, the composite panels are structural, so, for the coupé version, the A-pillar, the body sides and roof panel all contribute significantly to the torsional stiffness and structure, whereas on an Elise pretty much all the panel work is parasitic mass. So, in this structure, we have got uni-directional glass, high-strength glass-fibre running up the A-pillar into the roof structure, and in the body sides and windscreen surround, to give us improved torsional stiffness." Torsional stiffness rose from 14kNm/Deg to 26kNm/Deg, so the composite body structure nearly doubled the torsional stiffness. "That is with a glass-fibre composite, not an exotic carbon-fibre structure, which would be even stiffer, though obviously with a massive cost increase."

As well as static stiffness, there were the modal characteristics of the vehicle. Dave Marler explained: "As a very basic schoolboy example, if you put a ruler on the edge of a desk and twang the end of it, it goes off at a prescribed frequency, which is a function of its length and stiffness. In the same way, but obviously a bit more complicated, there are fundamental frequencies of a vehicle chassis. We have two key modes of torsion and bending. The bending mode of the vehicle is different to the static one, because it's a function of how the mass is distributed over the vehicle in terms of what frequency that it vibrates at. The modal characteristics are what you physically feel when you drive the car down the road, and as you go over a bump it tries to excite those modal characteristics of the vehicle. That has an impact on the NVH – noise, vibration, harshness – and the feel of the vehicle. These fundamental characteristics of the structure are tuned with optimisation for the chassis members and sill sections, to achieve the targets that we have set for the vehicle that will give the driving characteristics that we want.' The CAE Department was also involved in defining the strength and durability of components, ranging from suspension components to the adequacy of wing mirror stiffness when the door is shut. 'We had an issue when we changed the design of the rear spoiler between EP and VP, and the spoiler became too flexible, so we had to troubleshoot that. In a nutshell, the whole purpose of CAE is to try and get the design right first time, reduce the design time, minimise the risks so that no nasty surprises turn up in testing."

Richard Rackham – Vehicle Architect

The time-served and thoroughly proven Elise chassis was designed by Richard Rackham. That would be the pinnacle of many an engineer's career, but RR continued to apply similar thinking subsequently to a host of Lotuses and other Lotus Engineering commissions – obviously including the Evora. Before joining Lotus in 1986, Richard worked for Aldridge Engineering Consultants in Long Stratton, Norfolk, on the design of a Formula First racing car. His employer, Geoff Aldridge started off designing F1 cars for Team Lotus (along with Martin Ogilvie), including the Types 77, 78 and 79, and the consultancy was a major supplier of expertise to Lotus.

A graduate of Loughborough University, as a mechanical design engineer, Richard Rackham began his career within Vickers before working on a freelance consultancy basis. From 1987 to 1994, he provided design support within Roger Becker's Ride and Handling Department on numerous client projects including racing Esprits, the pinnacle of this being the 1993 Le Mans cars for Chamberlain Engineering. As well as the extruded, bonded and riveted aluminium chassis, the Elise amalgamates his experience in racing and ride and handling. He was responsible for design of the structure, suspension, controls and packaging, and patents for the extruded pedal and suspension upright design were granted as a result. Richard was then responsible for the design of the road-going homologation version of the V8-engined Elise-based GT1 race car (Type 115), which was followed in 1998 by the high technology structure of the Aston Martin Vanquish which was engineered at Lotus. Numerous client programmes and Esprit replacement studies led to the Evora era, and Richard was involved from the outset. He was responsible for packaging and conceptual structural design, with a brief to adapt his original Elise concept to suit the larger dimensions and characteristics of the Evora.

Given the permanent role of Vehicle Architect from 1999, two decades later, Richard oversaw the creation of the Evija and Emira.

Richard Rackham designed and instigated the extruded aluminium chassis, pioneered by Lotus in the Elise S1 and employed in every subsequent production car, including the Evora and Emira. (Courtesy Lotus Cars)

Forged aluminium

Lotuses have always handled superbly, partly because of their suspension setup, and the Evora proved no exception. The major elements in the suspension componentry were the forged aluminium wishbones, and the concept of a double-wishbone setup with anti-roll bars was carried over from the Elise. The forged aluminium material was a lightweight gain, since although the suspension arm on the Evora was necessarily able to cope with a heavier car than the Elise, by switching from steel to aluminium there was a compensatory weight saving. The forged aluminium component was more accurate dimensionally, because it had been machined, rather than a series of triangulated tubes welded together. It was therefore more able to absorb and withstand crash loads, and was also stiffer. It was space-efficient, because the aluminium could be shaped into a curve, whereas the steel tubes of the Elise had to be fabricated as straight sections.

The main investment came in the tooling, because initially that was more expensive than welding. So, to mitigate that cost, Lotus created a manufacturing process where a blank piece of aluminium was forged out, and the four different forgings went round the car to make all eight wishbones, since the same forging was employed on the left- and right-hand sides. This system had to be approved in the boardroom because of the extra investment needed, but it was deemed a necessary component, with payback over the Evora's lifecycle.

The Evora's bonded chassis provided accurate suspension pick-up points, tolerant to within half-a-millimetre, with thicker aluminium to spread the load. At the back-end, the welded steel subframe was necessary to disperse engine heat, since the bonded aluminium needed to be kept at below 100ºC to maintain the integrity of the bonding. The welded steel was constructed to do two things at once: to manage the impact loads in crash

Trelleborg – wishbone manufacturer

Based near Malmö in the south of Sweden, Trelleborg AB made the wishbones and bushes that are unique to the Evora. This branch of the industrial conglomerate specialised in bespoke solutions such as the Evora suspension for other automotive players, as well as manufacturing a selection of bushes and mounts, both conventional and hydraulic, applicable for vehicle platforms worldwide. Among Trelleborg's less exotic, but nonetheless invaluable products, were anti-roll bar bushes that acted as buffers between stabiliser mounts and the vehicle body, allowing the bar to be connected as stiffly as possible but also allowing it to move.

Trelleborg's subframe bushes connected the subframe structure to the vehicle chassis, isolating the body from structure-born noise and vibration transmitted by the engine. The company also made top strut mounts, designed to prevent sudden agitation of the car body through bumps and potholes. As well as damping, the strut mount also helped isolate tyre induced noise and vibration. Multi-link suspensions systems were becoming increasingly common in the motor industry, and rubber to metal bonded bushings were often used as joints between the actual links and the body. Trelleborg's input here included trailing arm bushes and front link arm bushes. It also produced boots to enclose CV joints to keep mud, water and salt out. Other innovations included friction-free or slippery rubber, while PU (polyurethane) elastomers were cost-effective alternatives to rubber in several applications, like bushing and bump-stops.

---

situations, sending them past the engine and up the sides of the car, and it was assembled so that the pivot points could be accurately controlled.

Working closely with suppliers such as Redaelli, COP, Trelleborg, Bilstein, Eibach, and TRW, the Suspension Design Engineering Team, including Mark Crowley and Carlo Besenzi, helped create efficient yet practical production solutions for the suspension and steering systems. The team reckoned that the rear subframe was the most challenging part of the suspension's conception, because there was so little space available for it, yet it still had to have a substantial load going through it. The high-powered Toyota V6 further reduced the space for the subframe, because Lotus was committed to minimising the overall width of the rear end.

Trelleborg AB, an industrial conglomerate based in the far south of Sweden, with its production unit in Germany, developed and made the suspension elements that were unique to the Evora. The idea behind the bush, in the same way as the Elise, was to achieve high radial rates, but with low torsional rate, thereby minimizing hysteresis and friction in the suspension system. This enabled better control of the suspension system through spring and damper tuning and tyre development, to obtain the best possible lateral stiffness in order to give drivers the steering connection and responsiveness so desirable in a sports car. The ideal bush component is a ball joint, but obviously that cannot be sensibly applied in a road car because the harshness factor would be too high. So, what Lotus tried to achieve with the bush was to retain the benefits of a ball joint, without its fragility and hard ride impact.

The stiffness of the suspension bushes could be upgraded by changing the rubber compound if Lotus or their customers wanted to increase the sharpness of the car. The team could go to Trelleborg with a specification for stiffer or softer rates. The work did not require much development, since the basics of the bush were correct, and with the bushes as well as the suspension and brake system, Lotus had a versatile platform that could be upgraded and tailored to suit most applications over the longer term. CAE ensured that the basic suspension characteristics would work on different vehicles, as and when required.

The bushes were the subject of much development, with at least ten progressive versions produced. The outside was rubber, but the interleaf – the element that allows for a high radial but low torsional rate – was aluminium. Trelleborg designed the interleaf in this way to meet the specification for stiffness demanded by its client, to meet the challenges of the targets for radial and torsional stiffness. It was therefore a bespoke component, and the key to achieving the Evora's exceptional steering, ride and handling performance.

CHAPTER 5

DRIVELINE
Powertrain selection and tuning

At launch the Evora's 3.5-litre Toyota-derived V6 was normally-aspirated, and developed 280bhp. It shared the engine with a number of top-of-the-range Toyota and Lexus models; the Camry is most often cited and, one suspects, perhaps with derisory intent. A motor out of a unprepossessing saloon car? The Land Cruiser SUV used a similar powertrain. There's a matter of pragmatism in engine choice unless you're a major producer: you want the most reliable bangs-for-buck you can get, and Toyota offered reliability, serviceability and horsepower. The Camry had used a 3.0-V6 since 2002, and when the Evora came out Toyota was producing a 4.0-litre version of its GR V6, begging the question why Lotus didn't fit that instead of going with forced induction in 2012? The Bolwell Nagari? Me neither, but this Australian sports-GT car

Toyota's 3.5-litre V6 promised reliability and power delivery commensurate with Lotus' aspirations for the Evora. (Courtesy Lotus Cars)

also used the same Toyota derived power unit from 2009. As fitted in the Evora, it's to a specification uniquely developed by Lotus.

It's a fascinating story, how the Evora's engine and transmission combination was arrived at. As we've seen, Richard Rackham's team took a generic V6 and used its dimensions to define the space needed to house it, taking into account cabin and boot spaces, wheelbase and track measurements. At the same time the quest for the definitive engine and transmission was on. The Executive Engineer of the Powertrain Controls Group, Paul Birch, and Powertrain Release Engineer Mark Ashby went through the powertrain selection, identifying which of the short-listed V6 engines it would be.

Paul described the process: "We made the decision to go with a V6 rather than four-cylinder turbo or V8, transverse-mounted, not longitudinal, and we then got the dimensions and the package detail from Toyota, and relayed that to Richard's team so that he could finalise chassis detailing." It didn't end there, of course. There were potential model evolutions to bear in mind. "We had to look at it and think, 'Okay, what are we going to be doing in the future?' That meant leaving space for a turbo or supercharger installation, and we made sure the engine bay would take the automatic transmission since that's physically larger and has additional coolers."

Since the chassis configuration was dependent on the shape of the transmission, Lotus had to decide which engine and gearbox to use, right at the beginning of the project. There had to be a degree of Toyota adaptation for the Lotus Evora application. It was a challenge to integrate a manual gearbox with the Toyota 2GR-FE V6, since it was originally intended to be mated to an automatic gearbox, so Lotus and Toyota worked together to identify a suitable manual example, and it was modified to get the correct baulk ring patterns, the starter-motor fixing points and the correct driveshaft angles for the Evora. The V6 was fed by a 57-litre fuel tank, similar to that of the Elise, made of stainless steel and located under the Evora's rear seats.

It was a juggling act with a lot of balls in the air. As Paul recalled, "Everybody is in this very fluid environment, and that early phase lasted two months, and the whole project was allocated 27 months total, so we couldn't spend longer on it. The styling guys were already working on themes and sketches, based on wheelbase and track measurements, and outlines of the interior which we'd given them, and everybody was trying to define their own bit of the car. Once we'd decided on the actual engine, we got detailed 3D CAD data from Toyota, which meant we could define a really big mechanical lump very early on. The speed of responses between ourselves and Toyota was partly due to the good relationship we'd built with it through the four-cylinder programmes, going onto the V6. Then it was down to us to work through the detail: where does each individual pipe run for oil, fuel or cooling go, and that takes a long time to establish. Then there's the development side; running the engines in test bays, testing the car in hot and cold places, going through extremes of temperature to make sure it operates correctly and gives the character we want from it as a Lotus. We are different from big car companies; Lotus leans on the characters involved, rather than the process and engineering, and that is partly what gives the car its character, because the people who created it put so much into it. People buy into a Lotus because of the company's history and its heritage, but the cars are quite unique because of the people involved in it. Their personality is in the finished product."

Of course, Lotus already had a multi-cylinder engine in its own backyard, and it raises the question, why not use the Esprit V8 in the Evora? According to Matt Becker, the V8 engine was never in the running. "That's a longitudinal layout," he pointed out, "and obviously the V6 is transverse. At one point they were actually making one up for an Evora with a longitudinal V6 in it, as a one-off, as a context for the future Esprit. So, they took the Evora platform, turned the engine around with a V6 transaxle transmission on it. But the V8 was never considered. The supercharged V6 was 350bhp, the V8 Esprit was 350bhp, so there was no power advantage in the heavier engine."

Tech spec – the Toyota 2GR-FE
Introduced in 2002, and designated the 2GR-FE, the four-main-bearing 3456cc Toyota V6's bore and stroke measure 94mm x 83mm. It has a 60° die-cast aluminium block and aluminium twin-cam cylinder heads. The engine's output varies according to what vehicle it's powering and what grade of fuel it's using, but is normally rated by Toyota at 276bhp at 6200rpm and 248lb/ft of torque. These figures are calculated when using premium 91 octane fuel at a compression ratio of 10:1.

The Evora's version featured Toyota's dual-VVT-I variable valve timing on both the intake and exhaust cams, and was managed by the Lotus T6e engine control unit. Valves were driven by roller-follower rocker arms, with low-friction roller bearings, with a unique, concave cam-lobe designed to increase valve-lift over the traditional shimless lifter-type system of Toyota's 4.0-litre 1GR-FE engine that powered various Land Cruiser derivatives. Overall cylinder head height was increased slightly to accommodate this slightly taller roller rocker system. The 2GR-FE cylinder heads were segmented into three parts: the valve cover, the camshaft sub-assembly housing, and cylinder head sub-assembly.

A 3.5-litre V6 engine already installed on the rear subframe, which is about to be attached to the central chassis module to the right. (Courtesy Lotus Cars)

It was a unique package, and an exclusive solution was needed to achieve it, as Paul Birch explained: "This V6 engine was only ever mated with an automatic transmission in Toyota vehicles, so because we wanted to have a manual gearbox for the car, Toyota offered us a six-speed manual transmission, transverse installation and correct torque capacity, and we did all the adaptation work here at Lotus." Also in the mix were a new flywheel and clutch assembly, new starter motor, and new clutch actuation systems. Putting it all together was a major feat of logistics and engineering, since that gearbox was never intended to be mated with that particular engine. "We kept Toyota fully informed all the way through, and a large part of the early programme was to ensure that we kept the integrity of their bits. We knew the engine was good, and we knew that at some point in the future we would want to produce an automatic transmission version of the car, but to establish its reputation, a pure sports car demanded manual transmission."

Designing and developing that interface, and mating the parts in the designated timescale was a huge achievement, but that was just the start. "We also developed our own internals for the gearbox, so we had a close-ratio gear-set to give the car more sporting character." So, there's a standard set of cogs and a sports set, and the original set was perfect for the automatic IPS transmission version of the car. "The first set is what came from Toyota, which was fine and it suited the car. With those, it achieved really good economy, and we got 205 grams per km CO_2 rating. But we then wanted to have something a bit more sporting."

Liaison between Lotus and Toyota came up with the appropriate transmission to handle the V6 engine's power output, sourced from a four-cylinder diesel driveline. (Courtesy Lotus Cars)

The new engine programme was totally secret, because it came at such an early stage in the Evora's development, long before any camouflaged cars were on the track or road. Even when I rode in the Esprit-clad Eagle mule at Bosch's Arjeplog site in northern Sweden, the engine's identity was not revealed, though as a six-cylinder connoisseur I was pretty sure that I could recognise the Toyota unit, or at any rate make an educated guess. At first, the engine was tested on the bench in a secure, confidential, static environment before being transferred into a mule car. Simon Hayward was the secrecy czar: "When the mule was up and running live, no one from Lotus wanted anyone else to know what it was. Last year, during southern European testing at Nardo and Idiada, we got away with it for several months before there were any EP camouflaged cars out and about." Famously, a Lotus Director went to see the mule, leaned against it and said, "OK, I want to see the real car, where is it?"

– the secret was so well kept that the mule looked just like an existing product.

Gearbox
After the review process and selection that finally pinpointed the Toyota 2GR, then came the 3D models and the adaptation of a manual transmission. "The V6 engine we chose didn't have an automatic transmission," reflected Paul, "so we adapted another Toyota gearbox, of which there was only one available that could withstand the torque, and that came from a diesel Avensis. Toyota obviously have some input into what they sell us, so we don't over-torque the gearbox, and they are quite stringent that the gearbox we use is suitable for our application. In this instance, it's not such a different torque output from a petrol V6 to a small four-cylinder diesel, so that's why the basic Evora drives very well, and is also very economical and yet still does 0 to 60mph in 4.9sec."

Selection of the RAV-4 engine had just as much to do with the installation and mounting brackets that enabled placement in the Lotus chassis. "You had to think beyond 'this came out of a pickup,'" Paul pointed out. "You had to think, 'it's the right capacity, it has the correct installation angle, the correct brackets,' and then we got on with the development. The standard RAV-4 unit produced 260-270 horsepower, we were producing 280PS, so our output was higher than any Toyota installation, though it was the same engine. It's the exhausts and the induction system that are different, and we optimised those in our installation, though we did use the original Toyota exhaust manifolds with their primary catalyst. But from thereon, the down-pipes, the main under-floor catalyst and the silencer were all bespoke for the mid-engined installation.

"Symbiosis is limited, though Toyota's senior management approves. They told us, 'You are doing a great thing there, you're showing our technology off in a really exciting way,' and the concept of what we were doing was filtering back to them. But they're a huge organisation with maybe 14,000 powertrain engineers, while we have probably got a couple of hundred. It's a different scale, but we hope in a small way some of that filters back to Toyota, and maybe it will help them move from their very worthy range of transport into some more exciting things in the future."

The first set of ratios was what came in the gearbox in the first place. Paul expanded on the scenario: "Bearing in mind their original application was in a Toyota, once we got cars running, we could do some serious analysis, and there was a lot of subjective and quantitive assessment. We decided we wanted to fit different ratios, and this was a complex three-shaft

---

Tech spec – Toyota EA60 gearbox
Manual 6-speed transmission; Toyota designation: EA60.
High-torque manual transmission, with twin final drives to maximise usage of package space and reduce gearbox length in a transverse installation.

Standard transmission ratios		Overall gearing
1st	3.538	(5.6mph/1000rpm)
2nd	1.913	(10.4mph/1000rpm)
3rd	1.218	(16.3mph/1000rpm)
4th	0.860	(23.1mph/1000rpm)
5th	0.790	(29.4mph/1000rpm)
6th	0.638	(36.4mph/1000rpm)
Reverse:	3.831	

Final drive: 1st-4th (including reverse) – 3.777
Final drive: 5th-6th – 3.238

---

gearbox, like a very compact transverse unit, so there was a fair amount of engineering work to install the different ratios. Those went into production, and that was another achievement: with Toyota's knowledge and approval we changed the internals of a Toyota gearbox to suit our application, and that underlines once again the close relationship we have with Toyota. They have the confidence in us to make such major changes, and are still happy to have their logo on the box. They recognise we've got the solid engineering background to implement this, and we understand what we want to do with our car to make it more exciting for the customer. That's why there are two ratio sets: the sports ratio is an option for those who want just a little bit more out of the car."

Automatic acceptance
The automatic IPS version launched in 2010. "It's an everyday car," said Paul, "aimed especially at markets like Japan and the States, where around 85 per cent of new cars are automatic, and even Germany where 65 per cent of all sports cars have just the two pedals. For that, we use the original six-speed automatic transmission, but with our electronics on top we can get the character and the installation correct, so that the automatic suits the sports car rather than a sedan." It was really important to Lotus that it got the electronic character right, so, clearly, it didn't just take an engine from some Toyota saloon car, drop it in and hope for the best. For instance, Toyota engineers who were intimate with the engine from its original applications drove the Evora and demanded to know, "What have you done inside the

The original console-mounted IPS push-button controls for the automatic shift feature conventional R, N, D and P symbols. (Courtesy Antony Fraser)

Controls Group and allowed it to apply its own engine controls and engine computer, so that they could characterise the way the engine behaved in the car electronically, as well as being able to interface it with all the vehicle's other electronic systems. The driving experience was intimately connected to the car's electronic characteristics, so Lotus had to control this aspect of the car's development. Manufacturers who farm out their engines and insist on controlling how it's deployed in the car create a climate where, effectively, a niche maker like Lotus is building one of its engine supplier's cars, to its specifications, and simply cladding it in the smaller company's bodyshell. And so it goes on, because if, for instance, the small volume player has to use a major manufacturer's engine computer, that means taking its immobiliser, ABS, body computer, dashboard, until the unique benefits of buying a niche model have been totally undermined.

This is precisely the predicament that Lotus was able to avoid, thanks to the flexibility of its Toyota deal. "The fact that we could fit our own electronics was really key to integrating the Toyota V6 with the car and getting the character we wanted out of it," Paul pointed out. "We can optimise the efficiency, and we can integrate with body systems, the dashboard, the alarm systems, which you couldn't do if you bought an engine package from another manufacturer. Generally, they would say, 'This is the engine, these are the electronics, and these are the systems around the electronics, and you have to use these,' and the Type 121 Europa was a classic example of that; GM said we could use their engine, but it came with their electronics and immobiliser, and so on, and that's all we could use. So, you have compromises because of that. It is still a very good car, but it could have been better if we'd had a little bit more autonomy with the way we could tune and adapt the engine by changing the electronics."

Before implementing electronic development, calibration and powertrain controls were addressed in detail, with masses of design activity and integration to ensure that the Toyota V6 fitted the chassis, the engine mounts were totally sound and accurate, and ancillaries like the exhaust system, air filter, cooling and oil systems would fit. A core member of the Evora project team from the start was Mark Ashby, Powertrain Release Engineer. He was responsible for making sure all parts arrived on site, matched and fitted, and were delivered into the project on schedule. His task was to drive the designers to get engine mounts, cooling pipes and exhaust systems mounted correctly, so he and Paul Birch were the principal technical liaison officers with Toyota, which meant making regular trips to Japan. Mark was also instrumental in mating the manual gearbox to the engine, motivating the powertrain design team and working

engine?" Paul Birch's response was, "Well, actually, nothing," and the Toyota techies took some convincing, mistakenly perceiving that some quite fundamental changes had been made. As Paul said, "People that knew the engine well thought internal changes had been made, but that was because of the electronic character we'd given it, the way it interacted with the dynamics of the vehicle. In fact, that was an interesting piece of feedback: the parents of the engine didn't recognise their child anymore."

Lotus was able to hone the power of the V6 specifically for the Evora's dimensions, while Toyota supported the Powertrain

Cockpit of an IPS Evora: steering wheel buttons and console buttons to select Drive, Reverse, Neutral or Park. (Courtesy Antony Fraser)

with suppliers to ensure everything fell into place. He was, in a sense, the architect, employing various detailed designers and draftsmen.

Paul elucidated: "Mark dealt with the componentry, making sure it all bolted together correctly, fitted on the production line, and capable of surviving the lifetime of the car." Paul's team was tasked with making sure that the powertrain remained reliable, and temperatures, pressures and speeds stayed within their design parameters, and within that, the character of the powertrain suited the rest of the vehicle. Mark dealt with non-Toyota parts, like the cooling and oil hoses that fit the Lotus subframe but are radically different to the Toyota installation. "It doesn't just arrive on a crane and you bolt in the back of the car; you've got to connect it up with all sorts of micro systems, and ensure all its vital fluids are flowing around in the correct way."

Because the Evora engine was mounted amidships, the

Mounted amidships, the Evora engine demanded different cooling arrangements to Toyota's original front-of-chassis location. (Courtesy Lotus Cars)

cooling arrangements were somewhat different to the original Toyota front-engine setup, throwing up big issues with the water-cooling system, the oil cooling and the component temperatures. "During the cooling tests we were looking at water flow," remembered Paul, "because the radiators were all at the front of the car and we had a huge number of pipes going round, which also affected heater performance and cooling. If you imagine any typical front-wheel drive Toyota, the cooling hoses are relatively short, whereas we've got two metres between the radiators and the engine. All the systems had to be reviewed, and that was Mark's expertise."

Sourcing the perfect components inevitably led to a melange of stock Toyota and bespoke parts. As Mark revealed, "We'd go through most of Toyota's parts bins to select the most appropriate items, and obviously things like the fuel pump is from Toyota, integrated into our own fuel tank. Some of the water tanks we used were from Toyota, but the Evora's radiator and the hoses were all bespoke, and totally different and unique to suit our installation. That meant working very closely with the chassis group, because the engine mounts and the subframes was obviously a crossover between my function and those of Ingemar Johansson and Clive Birchall."

Lotus also developed the Evora's clutch and flywheel and pressure plate assembly, because nothing appropriate for a manual installation was available from Toyota. The assembly was made by AP Driveline Technologies in Leamington Spa. Being an engineering consultancy in its own right, Lotus is ideally placed to design engine transmissions from a blank sheet

Surmounted by an Edelbrock supercharger, the 3.5-litre Toyota-derived V6 in this GT430 develops 430bhp. (Courtesy Lotus Cars)

of paper, which it does for other companies and, indeed, had done for the Esprit V8. "We don't do that for the current cars, just because of the volume," Paul clarified. "We could design and develop it, but to produce it doesn't make economic sense, so that's why we partner with Toyota, because they have higher volumes and good technology, but when there are bits where we need to pull it together, as a consultancy, we can work from first principals; we have the experience and the background to make the calculations, the designs, the expertise to take it to production."

Powertrain calibration

Powertrain calibration is a Lotus shorthand term used to describe the Powertrain Controls Group's integration of everything that makes the engine work. That incorporates the ECU – engine control system – and all the development surrounding its interaction with the other systems in the car. As well as Paul Birch and Mark Ashby, another key figure on the Evora driveline was Simon Hayward, Senior Engineer in the group, who led the team on tests for the integration controls. Simon had been at Lotus for almost 20 years at that point, having started on the production line building Esprits, Excels and the M-100 front-drive Elan. Following an engineering apprenticeship he eventually moved into Paul's group in 2000 to work on the Series 2 Elise. "There were between eight and ten people working on powertrain controls across the whole Lotus range," Paul explained, "so there was plenty of continuity, which helped retain that distinctive Lotus family resemblance throughout the models, up to and including the Evora. Everyone working on the Evora, throughout the company, knew the kind of car they wanted to produce, and the experienced engineers on the team brought with them a unique combination of knowledge about the legislative and consumer requirements for a modern car, with enthusiasm for the pleasures of driving that are so fundamental to the Lotus character."

The relationship between Lotus and Toyota continues to be a symbiotic rapport, allowing a two-way dialogue. There is a level of mutual respect and learning, from which both parties benefit. Historically, Toyota's support with technical information and its open lines of communication – going back to 1983 when Toyota supported Lotus at a corporate level in the wake of Colin Chapman's death. This made contact fairly straightforward, since any request for information received a quick and full response. Paul's association with Toyota dated from the switch from Rover K-series to Toyota power for the Elise and Exige family in 2004, so for him, the Toyota 2GR-FE V6 was very much a preferred option from early on in the project. All V6 units potentially available were checked out at concept stage, but many were rejected on the grounds of performance, weight or cost, not to mention the improbability of suitable supply arrangements. In the end, there were very few V6s in competition with the Toyota version, and it proved an excellent choice. The Powertrain Controls Group developed its own ECU for the Evora, because it had those skills in house, and it also configured an ECU to operate with the automatic gearbox, so there would be no timescale difficulties.

Paul Birch's involvement began when Lotus started discussing the Eagle project. "The Powertrain Controls Group will be sticking with the Evora for as long as it is produced (which turned out to be 2021), including after-sales." He was looking that far ahead from the outset, and back in 2007/8, his team was still dealing with the Esprit V8 and the Elan M100. So, that was some commitment. As he pointed out, "Customers still drive them, and the company retains its obligations to look after the cars in the field. Lotus also has a legal obligation from the point of view of emissions legislation and warranty perspectives." Powertrain control is an encompassing discipline that continues long after a car is forgotten by most people at the factory, apart from the ones supplying spares.

For the Powertrain Controls Group, the start of the process was the arrival of a consignment of Toyota engines on the Hethel site. First step was for Simon Hayward and the team to put the engine on the dynamometer, a semi-state test bench, with the basic building blocks of the calibration software, to create a map of the engine, its performance output, emissions and fuel requirements. The engine had electronically-controlled direct ignition, and quad VVT variable cam timing on all four camshafts, and variable induction system lengths; these variables were modified on the dyno to obtain the best power output, the lowest temperature exhaust, the optimum fuel consumption and the lowest emissions for the exhaust in each operating zone. In a rigorous process lasting several months, the engine was profiled and modified at low engine speed and throttle position, in low acceleration position, with good emissions and fuel economy, and also at high revs when the throttle is wide open and maximum power output is required.

Once the engine was mapped and modified, the Powertrain Controls Group designed its characteristic electronics, wrote the software for the controls, and developed all the calibration – an exceptional thing for any manufacturer to undertake in-house, let alone a car maker of the size of Lotus. "Nevertheless," said Simon, "it was essential to give the Evora that unique Lotus flavour. A volume manufacturer would outsource that work to Bosch or Siemens, for instance, beginning with the electronics and using them to drive the design of the engine." Lotus' standpoint was different. It started with the engine and developed the electronics around it as required by the profile of the new car. That degree of flexibility and functionality was now the norm at Lotus. It began with the Esprit V8, when no one else could be found to supply an electronics system to suit the company's needs. Lotus had worked on engine control systems since they were invented in the early 1980s, and it built up exceptional in-house expertise, consolidated by Lotus Engineering's consultancy projects it undertakes for third parties, which are largely secret and often involve new technological developments. According to Simon, "The trick is to combine the best possible ABS system – for instance – and electronically interface it with a dashboard that fits both the styling and the functionality of the car, and then to interface that with Lotus electronics. So, by having its own electronic expertise, Lotus retains the flexibility to pick and choose hard- and software to suit the character of the car." Paul related the virtual world of electronic management to the physical world of chassis design: "The character of the car has to be seen in the electronics, just as much as in the chassis, for it to be a real Lotus. Just as the ride and handling derive from the chassis, the suspension system and the choice of tyres, we build the same flexibility into the electronic architecture of the car."

With the Evora being effectively entirely new, the powertrain team had to focus on the integration of a series of new concepts: the new subframes and rear suspension system, packaging for the air-conditioning systems, all the incoming electronic systems, a new dashboard, and new ABS configuration. So, the actual

As Evora chassis pass down the production line, the pre-assembled powertrain is hoisted into place. (Courtesy Lotus Cars)

engine management component was a relatively minor element of the work. The ABS development programme was carried out in association with Bosch, and involved a huge amount of work for the Powertrain team on account of the interaction between the ABS and traction control systems and the Bosch system. Lotus engineers developed the interface between the two, so were in regular communication with Bosch vehicle dynamics experts, visiting each other whenever there was a software update from either party, on average at least once a month. Bosch normally expects software changes from major engine control manufacturers to take months, but they were getting Lotus feedback in hours, since data and specification recommendations were written in-house at Hethel by Lotus' own software authors as well as out-and-about on test trips.

Mule team
Going step by step, the Powertrain Controls Group took the modified engine and its associated controls through design,

testing and validation in the workshop. It then saw how it all came together in an actual vehicle – the mule – before the company committed to the EP development programme. The group worked within the test cell at Hethel to mimic possible exhaust, airflow and induction systems as accurately as possible. Vehicles on the Eagle programme, morphing into the Evora, ran for eight months before there was a ride-and-handling car, or a cooling car to house them in, and before either of these there were also two powertrain development mules on the stocks. Among these was a RAV4 Toyota which the team used as a test platform, a bizarre disguise for a sports car engine if ever there was one, being a dumpy 4x4 SUV. It was fitted with the V6 motor in its new modified configuration, and was a sound choice as a mule because it had the advantages of housing the engine in its native environment, connected to a set of proscribed performance rules, and held as if in a bracket off the ground. It was subjected to the full gamut of tests, exhaust emissions, fuel consumption and high ambient testing. Lotus moved early to get the powertrain into a development vehicle, because it had to be ready before cooling and other systems could be incorporated for testing, and that pre-dated any experimental prototypes being built.

The Evora's exhaust system was made of stainless steel to a Lotus design, by the Derbyshire-based firm CLF. It incorporated elements of the Toyota system, which helped to control material costs, since Lotus hadn't had to design and manufacture them from scratch. Its design had to comply with the legal noise test standard, 74dBA – decibels – drive-by for a two-gear test, and 75dBA for a one-gear test but, as the Evora was a sports car, the exhaust still needed to sound suitably fruity. Lotus then had to tune the pipe lengths and bores to reach a back-pressure target. Power-sapping as it is, the catalytic convertor actually helped with noise reduction, and Paul estimated it was worth about 1dBA. "The fruity-or-legal argument was a difficult one to square," he mused, "particularly with wide tyres, which made a major contribution to noise levels in the high-frequency noise range." Having driven several electric Porsches including the eRuf and Ruf Greenster, which were reasonably quiet apart from the high-pitched whine, the substantial road noise generated by their tyres was a real ear-opener, and something we're mostly oblivious to, due to engine noise, air-con, hi-fi, and so on.

The standards for the noise tests were extremely restrictive and hard to meet. Weather played a part, too, since temperature drops make the air density rise, making it harder to meet the dBA standard. So, a warm ambient temperature helps, since there is no correction factor in the test, which is why most manufacturers head south to Spain to have their cars certified. The tests cannot be performed in wet or windy weather, and have to take place on an ISO-accredited site. The Evora was certified for Europe at Idiada in Spain, which entailed a 1140-mile round trip to and from Barcelona. It was certified for Japan at RDW (Rijksdienst voor het Wegverkeer) in Holland, the Netherlands' national authority for road traffic, transport and vehicle administration.

A major aspect of Paul Birch's brief was NVH – noise, vibration and harshness – management, the art of refining the cabin environment to make it a pleasant place to be by dealing with issues like wind and pass-by noise, tyre drumming, and intake and exhaust development. Leading the team on interior refinement was Matt Barr, reducing NVH by addressing all the sealing issues, checking all the joints and the gaskets around the doors and windows. After sealing, acoustic treatments were applied, including PVA damping sheets formed from a tar-like product called DFS, which were added to panels as liners. Once again it was a trade-off between performance and weight, since you cannot simply fill all the car's cavities with foam and expect to maintain appropriate performance.

Testing started at Hethel's on-site track, with computer communication set up inside the mule cars via laptop and a set of electronic calibration tools. The test cars were also taken to Millbrook, Bedfordshire, Europe's leading commercial test and validation centre for the automotive industry, which has high-speed banking and cobbled pavé to really make the vehicles (and drivers) suffer. Then they were subjected to extreme climate tests, driven to the Arctic Circle (Sweden's Lapland), the desert (southern Spain) and to Andorra in the Pyrenees, Nardo test track in southern Italy, and the Austrian Alps. As I shall shortly demonstrate, the Evora worked very well in extremes of temperature, hot and cold. Cold weather testing on the frozen lake at Bosch's Arjeplog test centre showed that it was still driveable at minus 20 degrees. Each trip lasted two to three weeks.

"This was no transporter job either," affirmed Paul. "The cars were driven to and from each of the test locations, disguised as mules or camouflaged prototypes, to build up the mileage and offer plenty of opportunities for adjustment, checking and recalculating the calibrations. The team members were all petrolheads, and they actively wanted the Evora to be a pleasure to use; comfortable but sharp, a proper sports car. It was a constant process of evaluation, assessing what the car felt like on motorways, through towns and built-up areas, how involving it was for the test driver, just as it should be for the customer." For Paul and Simon, the most difficult thing to project-manage was the supply of components in the aggressive timescale imposed right from the beginning – just 26 months. For example, the

The mule sent to the Bosch Arctic proving ground at Arjeplog was disguised within a stretched Esprit bodyshell. (Courtesy Jason Parnell)

original designated exhaust system manufacturer was taken over, and fresh arrangements had to be made. "Things like that can throw a project out of kilter, unless remedial action is taken very fast," said Paul, "so the availability of parts, designed and manufactured to the correct quality level demanded considerable thought." According to Paul, "It can be difficult to operate with a new supplier, as has been the case with the exhaust system, the driveshaft manufacturer, and the ABS system, where it takes time to establish relationships and communication paths. We managed to achieve that swiftly enough, though."

On such a fast-moving programme, it wasn't sufficient just to build a couple of mule cars in-house; Lotus needed more than 20 identical cars for validation, and all their components had to be to the correct standard, and assembled to roadworthy quality by suitably qualified technicians and fabricators. "It was a massive task, and meant finding a way round complex logistical problems," Paul recalled. "In practice, I would say to Simon, 'You're going to Nardo at the end of August for the final hot weather test to validate all the components,' and he would say, 'But I haven't even got half a car ready, I haven't got all my components.' But there was no choice; he would have to get hold of them, whatever the issues. There'd be no point going to southern Italy or Spain in the winter, or to Lapland in summer."

The fact that they were able to deliver the car to these vital schedules testifies yet again to the power of a small, flexible, highly experienced and committed team of multi-skilled engineers, many of whom had worked together over many years and numerous development programmes to achieve a high level of efficiency.

Watching America

As well as the ride and handling, the engine and transmission were also honed in countless hours of road and track testing. The powertrain needed to be evaluated in extreme environments too, and Paul Birch recounted a drive in the USA: "We had already launched the European car, and we'd done a lot of testing in Europe for the Federal car, so what we were evaluating in the States were specific differences in territorial requirements. Different fuel grades, different driving conditions, different climates, too. You could easily spend the whole day on one Interstate and never make a turn."

I can vouch for this, having motored an Elise S from Indianapolis to Daytona, covering almost 1000 miles in 19 hours, in a bid to link America's two major racetracks, both of which laid claim to being 'the World's Auto Racing Capital.' It reminds you that this is a big country. As Paul said, "I spent two-and-a-half days on one road, having pulled out of my hotel car park in the morning, and the satnav said 637 miles to the next hotel without making a turn. We went down to Texas, where it was 48 degrees, incredibly hot, and then we did high-altitude testing in Colorado,

A trio of Evoras on a 2009 media outing pause for the mandatory snapshot beside the old pits complex at Reims. (Courtesy Antony Fraser)

Unsurprisingly, as I found when motoring from Indianapolis to Daytona with an Elise, much of the Interstate Highway is interminably straight. (Courtesy Jason Parnell)

which you can't get in Europe. You can't find those altitudes even in the Alps, where the maximum you can get is 3000 metres, whereas Denver is already at 4500 metres before you climb into the Rockies. Matt Becker came out with us to check the ride conditions: because the roads are different everywhere, there are different expansion joints and frost heave, and some of the roads are horrendous, though, to be fair, some are really good. So, we did a number of certification tests at altitude just so we had all the correct testing done to make the formal application for the USA."

Another key issue that was evaluated in the States was the Evora's air-conditioning, which was cause for concern when I drove one to northern Italy, and confirmed by the powertrain crew in North America."We found that, when it was really hot, we needed better air-conditioning," said Paul Birch. "You go out there to test the engine, and you end up re-working the air-conditioning system, and that's another advantage of being a small company; you roll your sleeves up and get on with it, sending information back to the project team. We were living with the car like a customer. Most of us are car enthusiasts as well, the same as our customers, so living with the car as a customer is very important. Rather than just sitting behind our screens and doing some CAD and ticking our reports, we live with the cars. We want them to be good in all aspects, so we go and live with the car for weeks." Of the team that was in the States, three of the guys were out there for six weeks. Simon Haywood led the team with a technician alongside him, and the other members joined him as necessary to assess the heat, altitude and urban cycle. "It's a really important aspect of a car's validation, to live with it, not just to tick some boxes but to actually go out on a road trip and drive it."

Superchargers
Sure, and it wasn't long before the Evora gained the supercharger that levelled its performance with contemporaries like the Porsche 911, raising output from 275bhp to 345bhp at 7000rpm. This was in 2010, and the model appropriately gained an S suffix. In fact, the Evora's superior handling balance gave it the win by a nose. The 'blower' was an Eaton TVS (twin vortex series) supercharger, its installation engineered in conjunction with Australian firm Harrop, and as well as lifting power it also increased torque by 35lb/ft across the rev range, peaking at 295lb/ft at 4500rpm. There was none of the rasping whine that characterises the supercharged Exige, and the instantaneous

throttle response and linearity of the power delivery gave the impression that there was simply a bigger engine in the back and, if the supercharger wasn't visible in the rear window the casual observer might not have guessed that the Evora had forced induction at all. It was cooled by a pair of oil-water radiators mounted in the nose, along with the associated pipes and, along the lines of 'there's no such thing as a free lunch,' the whole installation added 50kg to the kerb weight. A side effect was that the car's shock absorbers were recalibrated in support.

Like night follows day, pretty soon independent Lotus Specialists such as Hangar111, Hofmann's and Monkey Wrench Racing were offering aftermarket superchargers such as Komo-Tec and Harrop alternatives.

Special editions
The Evora Cup GT4

Bearing in mind the company's motorsport heritage, it was a given that the Evora would be shaped into a competition car, sooner or later. Indeed, the Evora Cup car made its first appearance at the Autosport Show at the NEC in January 2010, plus the intention to run a points-based European one-make series. The Evora Cup GT4's factory type number 124 coincided neatly with the intention to enter it in endurance races such as the Le Mans 24-Hours. A further motivation was to endow the new model with some instant credibility as a sports-racing car, and the rigours of long-distance endurance racing with its twin goals of reliability and efficiency would see to that. For its first season it would run as a factory-supported car in GT endurance events, with customer versions available for 2011.

Visually, it was an Evora in aftermarket boy-racer mode. Based on a Dallara aero package, all the racing modifications were integrated in the Design Studio: Kevlar door panels, swollen side sills, front splitter, rear bumper assembly with diffuser, and louvered engine cover, finished in Epsom Green with yellow stripes, harking back to Team Lotus' pre-sponsorship era – rather like the Elise Type 25 'Jim Clark' model.

The Evora Cup GT4 at Snetterton for the 2010 Lotus Festival. It was powered by the supercharged 4.0-litre Toyota V6 with Cosworth electronic engine management, developing 395bhp. (Courtesy Johnny Tipler)

An Evora Cup GT4 with Martin Donnelly and Gavan Kershaw sharing the wheel leads a 2-Eleven and 340R at Snetterton, 2010. (Courtesy Johnny Tipler)

The intention was to enter the production GT classes based on the road car and not a silhouette racer. All the cabin trim was removed, as you'd expect in a race car, but the dash panel was retained. A T45 steel rollcage was installed, and to comply with safety regulations a fire extinguisher system was plumbed in, the battery box located in the passenger footwell, Recaro race seat, harnesses, while the rear-three-quarter side windows morphed into fuel filler apertures, and the tank had a slightly larger capacity for endurance racing. Aerodynamic additions such as the front splitter, rear diffusers and wing were defined

The Evora Cup GT4 imitated the livery of the Jim Clark special edition of the 2008 Elise, pictured here on my road trip to Portmeirion. (Courtesy Jason Parnell)

The Evora GT4 was a successful contender in the 2011 British GT Championship, driven by Ollie Jackson and Jack Drinkall. (Courtesy Johnny Tipler)

Resplendent in red, white and blue, the Evora GT4's Pollock-esque livery promotes Anglo-Japanese synergies. (Courtesy Lotus Cars)

Tokai ran an Evora for five seasons in the GT300 class of Japan's Super GT, with backing from Lotus Cars Japan. (Courtesy Lotus Cars)

Stanbridge Motorsport ran James Simmons and Fraser Smart in an Evora GT4 in the 2019 and 2020 British GT Cup series. (Courtesy Sarah Hall)

by the regulations, but carefully integrated so they suited the lines of the car. The powertrain was a 4.0-litre, 395bhp version of the Evora's normal 3.5-litre normally-aspirated engine, with Cosworth electronic management, allied to a Cima 6-speed sequential gearbox. Maintaining its relationship with the road car, the GT4 used the same suspension with forged aluminium double wishbones, Eibach springs and dampers, plus AP Racing brakes. The exhaust was unrestricted but retained the catalyst. The Cup GT4 ran on Pirelli slick tyres mounted on 18-inch centre-lock Rimstock wheels. The fronts were the same size as the road car's, but the rears were smaller in diameter than the road car to suit the characteristics of the slicks. At the time, it

The Evora Cup GT4 bombing around the Hethel test track in 2011. (Courtesy Lotus Cars)

was claimed to be the most powerful road-going Lotus ever, hitting 62mph in 4.0 seconds and maxing out at 167mph. Track width was increased by 110mm, and the GTE was also equipped as standard with a robotised paddle-shift version of Lotus' Aisin six-speed manual gearbox.

Developed by Lotus Motorsport, the Evora GT4/GTS retained the S's supercharged 3.5-litre Toyota V6, with six examples built. Lotus entered two Evoras for the 2011 Le Mans 24-Hours, run by JetAlliance Racing. Despite overheating issues during practice and qualifying, #65 finished 22nd overall completing 295 laps, though #64 retired after 126 laps. In 2012, Alex Job Racing ran an LM GTE in the American Le Mans Series, though with no success. Other notable outings included Olly Hancock winning the Nürburgring round of the GT4 European Cup, while Johnny Mowlem, Stefano D'Aste and Gianni Giudici finished on the podium in the 2011 Dubai 24-Hours.

In 2011, Stefano D'Aste led the GT4 European Series till the last race, when an engine problem dropped him to 3rd, which has to be seen as something of a result, considering it was the first year that the Evora GT4 did a full season competing with BMW M3s Porsche 911s and Aston Martin Vantage GT4s. In 2012, Richard Adams David Green and Martin Byford won the Britcar MSA British Endurance Championship in the Team Bullrun Evora GT4. And in Japan, Tokai ran an Evora for five seasons in the GT300 class of Super GT with backing from Lotus Cars Japan. Engineered by Mooncraft, the Evora GT300 MC was built to the series' Mother Chassis regulations and, as such, was a silhouette racing car, built on a Dome chassis and powered by a Nissan VK45DE naturally aspirated V8, and sharing almost no parts whatsoever with its production counterpart. Nevertheless, it looked the part, and scored its maiden victory at Fuji Speedway in 2020. Lotus Motorsport built 30 examples of the Evora Cup GT4.

A Type 124 Evora GTS Enduro on the Goodwood Hill at the 2012 Festival of Speed. (Courtesy Antony Fraser)

Evora Enduro

Lotus unveiled the Type 124 Evora Enduro GT at the 2011 Geneva Show, alongside the stripped-out Elise Club Racer. The Enduro was based on the Evora GT4 Endurance race car. It was shown in black and gold with splashes of red, just like the contemporary Lotus-Renault F1 car, which was also present at the Lotus stand – and although channelling Team Lotus livery from the JPS era there was no connection with the former Team Lotus, nor Classic Team Lotus.

Developed by Lotus Motorsport, the Enduro was the base model from which GT2/GTE homologated cars for FIA and ACO endurance racing were built, with plans to compete in the Nürburgring ADAC 24-Hours, Spa 24-Hours, and Silverstone 24-Hours as well as Dubai and Daytona. The Enduro was powered by the normally-aspirated 4.0-litre V6 with Cosworth engine management, developing 440bhp and paired with an XTRAC 426 six-speed sequential gearbox. Fully compliant with FIA regs in terms of fuel system, driver seat, roll-cage, the Enduro was equipped with Ohlins dampers, uni-ball wishbone joints, Alcon six- and four-piston brake calipers front and rear, forged aluminium alloy wheels, and an on-board four-point jacking system for quick pit stops. Unladen weight is 1190kg. Just six cars were built.

As driven by James Rossiter, Johnny Mowlem and Jonathan Hirschi for JetAlliance Racing at Le Mans in 2011 in the LMGTE Pro Class, the 4.0-litre Evora gets a ragging on the Hethel test track. (Courtesy Lotus Cars)

The winning Evora GT4 Enduro of Martin Donnelly and Gavan Kershaw in the Snetterton pitlane during a round of the 2011 Lotus Cup UK series – Martin's first race in a Lotus since his stint in the Type 102 in 1990. (Courtesy Johnny Tipler)

The Evora Cup GT4 burns rubber on the Hethel test track. The rear wing and diffuser are prominent additions. (Courtesy Lotus Cars)

A Type 124 Evora Enduro GT 4.0-litre V6 driven by James Rossiter up Goodwood Hill at the 2011 Festival of Speed. (Courtesy Antony Fraser)

The limited-edition Evora GTE from 2011 featured lightweight panels, and was powered by the supercharged 3.5-litre Toyota V6 VVT-I, producing 444bhp. (Courtesy Lotus Cars)

Evora GTE

Following on from the GTE Race car, the GTE was offered as a road car and, ensuing from exposure at the Monterey Jet Center in 2011, Lotus received 115 orders. Carbon-fibre panelling was deployed throughout much of the car, and its supercharged Toyota V6 produced 444bhp via an AMT racing transmission. All-up weight was 2623lb (1190kg), making it 230lb lighter than the base model Evora. The GTE was unveiled in September 2011 at the Frankfurt Show.

The Evora 414E Hybrid

Lotus thinking has always been at the forefront of technological innovation and, in 2011, a prototype hybrid Evora, designated the 414E, was unveiled at the Frankfurt Show. Developed as a demonstration project for the UK Government's Technology Strategy Board, uniting Lotus with Jaguar, Infiniti and a number of battery tech firms in the process, the copper-coloured coupé had an electric-only driving range of just 30 miles. Two 152kW electric motors (each delivering 204bhp) drove the rear wheels and, when combined with the 3-cylinder 1198cc Lotus Range Extender petrol engine, which acted as a generator, the 414E's range increased to 300 miles, emitting 55g/km CO_2. The 414E was no lightweight – always the issue with batteries – tipping the scales at 3878lb (1759kg). Still, it was trend-setting in 2011.

On display at the 2010 Goodwood Festival of Speed, the 455bhp Evora 414E hybrid demonstrated Lotus' forward thinking and technical prowess. (Courtesy Lotus Cars)

The Evora 414E hybrid used two 1.5kW electric motors mounted side-by-side, allied to a 1200cc 3-cylinder Lotus 'Range Extender' engine. The battery pack was mounted amidships where the plus-2 seats would normally be. (Courtesy Lotus Cars)

Evora Sports Racer

In 2013, those of a boy-racer persuasion were catered for with the introduction of the Sports Racer. Finished in one of four exclusive colours – Ardent Red, Carbon Grey, Nightfall Blue and Aspen White, with black roof and highlights, it was fully-loaded with all the top-line gizmos. it could be ordered with normally-aspirated 3.5 V6 or supercharged, though I don't know any boy racers – least of all me – with between £58- and £65-grand to splurge on a state of the art Evora. I'm being a tad facetious, of course, because any car tweaked by the factory that produced it is going to be an attractive proposition, and in this case worth every penny. The Evora Sports Racer was endowed as standard with a host of extras, including switchable sports mode with sharper throttle response, increased RPM limit, a sportier setting for dynamic performance management (DPM), plus rear diffuser and cross-drilled brake discs. The cabin was furnished with fabulous Venom Red leather sports seats, and gunmetal dashboard. Hi-fi and Bluetooth was comprehensive. This tasty limited-edition model was invoiced to 250 customers.

Fully loaded: the Evora Sports Racer, pictured here during the 2012 Goodwood Festival of Speed, was endowed with almost every conceivable option. (Courtesy Antony Fraser)

The 'James Bond' stickered Evora 410 Sport was acquired in 2020 by a Portuguese collector. (Courtesy Pedro Domingues)

The name's Bond; James Bond

The James Bond Evora was created in 2017 to celebrate the 40th anniversary of *The Spy Who Loved Me*, in which 007 drives off a pier into the Mediterranean Sea off Sardinia and his Esprit turns into a submarine. The suitably white-hued Evora tribute car was based on a 410 Sport, and bedecked with visuals honouring the movie-star Esprit, like the garish tartan interior and 007-specific Evora logo on the sides and rear, with the same font as used on the Esprit. Meanwhile, the Bond Esprit was on display in the Cars of the Stars museum in Keswick, and one of the stunt drivers in the movie was none other than Roger Becker.

The Bond Evora was presented as a one-off, but word is they actually made two cars: one that stayed at Lotus and another, produced for an Italian car collector who also has a Lotus dealership (I wonder who that could be?!), although this second one was kept a secret. The only visual difference between both these cars is the seats. The car revealed to the press had sportier seats, and the holes for inserting a four-point harness are white, while in the second one the touring-style seats are black and tan chequer. The official Bond Evora was acquired in 2021 by the current owner, a Portuguese collector, after a year of negotiations, on the assurance that the Evora would go to a good home and join a number of other vehicles already in the

Allegedly, two examples were produced of the James Bond Evora, based on the 410 Sport and created in 2017 to celebrate the 40th anniversary of The Spy Who Loved Me. (Courtesy Pedro Domingues)

Naomi Campbell and Lotus Cars created the 'Naomi for Haiti' special edition Evora, with all eight units auctioned and proceeds going to relief efforts in Haiti, devastated by an earthquake in January 2010. The first three cars fetched over $500k each. (Courtesy Lotus Cars)

new owner's collection. Since the Evora Sport 410 was never officially imported into Portugal, it's a rare sight on the streets of Porto. Submersible? My informant Pedro Domingues thinks not.

Blowing in the wind: from Harrop to Edelbrock

In 2015, Lotus upped the forced induction game by switching from Harrop to the US-made Edelbrock supercharger and installing an intercooler. This in turn lifted power once again, up to 400bhp, launched in the eponymous Evora 400. Acceleration from 0-60mph (97kph) took 4.1 seconds, half-a-second quicker than the out-going S model. Top speed was 186mph (299kph).

Replacing the Evora S in 2015, the Evora 400 was first in line to use the Edelbrock supercharger instead of the Harrop version, introducing an intercooler at the same time. (Courtesy Jason Parnell)

An Evora 400 captured on shakedown at Snetterton circuit. (Courtesy Stephanie Ewen)

Identifiable externally by its aerodynamically enhanced nose, featuring large air intakes on either side, and the rear wing became a split wing, contributing an additional 51lb (23kg) of downforce. The cabin interior was also slightly changed, with a new dashboard and centre-console, while the sills were narrowed for ease of access. Forged wheels were available as an option.

Evora Sport 410

The Evora 400 was succeeded by the similarly supercharged Sport 410, GT430, GT430 Sport and GT410 Sport, all of which appeared in quick succession, as though floodgates had opened and Lotus had at last found its feet again after the managerial hiatus of the preceding five years. Here's how they evolved.

In 2016, the Evora Sport 410 was released at the Geneva Show. The rear hatch/tailgate and font access panel were now in carbon-fibre, and performance was slightly increased by 10bhp, hence the 410 designation, while torque was raised to 310lb/ft. Lighter wheels and a lithium-ion battery were fitted, but possibly the most significant alteration was the replacement of the sculpted rear seats by one-piece carbon-fibre bucket seats. It could now do a supercar 190mph, and 0-60mph was knocked off in 3.9 seconds. There was always a slight performance penalty with IPS automatic transmission – though not in the case of Porsche's PDK shift – and the Evora with the IPS automatic loses out quite significantly, running out of steam at 174mph (280kph). Still, traditionally, automatic was always meant to be a more relaxed

The supercharged Sport 410 prefigured a rash of new Evora models, all with slight variations in designation, reflecting increments in power output. (Courtesy Lotus Cars)

Launched in 2017, the limited-edition GT430 (left) and GT430 Sport differed insofar as the former featured carbon-fibre aero panels, and the latter incorporated larger carbon-fibre air ducts and rear wing. (Courtesy Lotus Cars)

mode. And at that velocity, who's looking at the speedo anyway? Then, the Evora 50th Anniversary special edition was released to mark the company's Silver Jubilee half-century in 2016.

A year later, coinciding with the arrival at Hethel of new parent Geely, the Evora GT430 was launched. Figurehead of the new performance range, with just 60 cars built, the GT430 was the most powerful in the line-up, and the fastest street-legal Lotus ever made. Power output was now 430bhp (321kW; 436PS) with torque measuring 325lb/ft (441Nm) in manual transmission and 332lb/ft (450Nm) in auto, with launch control and quick-shift linkage delivered via its Edelbrock supercharger and titanium exhaust. It was also the lightest in the range, weighing 2773lb (1258kg) in manual format and 2888lb (1310kg) in IPS – a legacy of the automatic transmission's cooling system.

Appearance-wise, the GT430 sported a more aggressive bodykit, featuring carbon-fibre panels and a larger carbon-fibre rear wing, plus bigger frontal air intakes, polycarbonate lights and forged aluminium wheels. The GT430 also generated more downforce – 250kg at 190mph – than its predecessors. In the cockpit, the seats were now of the racing variety, and the rear ones were stripped out altogether. As well as carbon detailing, the passenger side dashboard was enhanced with a plaque bearing the car's production number.

This Tudor manor house, Ketteringham Hall, provides a backdrop for the limited edition Evora GT430. (Courtesy Lotus Cars)

Clockwise from above: Introduced in 2017, the GT430 was the most powerful in the line-up – the clue's in the designation – and the fastest street-legal Lotus ever made. (Courtesy Lotus Cars)

Cabin of the GT430 included monogrammed seats, swathes of Alcantara, and a plaque indicating its build number. (Courtesy Lotus Cars)

The Evora 50th Anniversary special edition was released to mark the company's Silver Jubilee in 2016. (Courtesy Lotus Cars)

The Evora GT410 Sport, snapped here on Cromer seafront promenade, was top of the range in 2021. (Courtesy Johnny Tipler)

Later in 2017, another variant of the Evora GT430 was introduced, the Evora GT430 Sport. It was clad in an even more aerodynamic bodyshell, featuring black carbon panelling, such as the Zagato-esque double-bubble roof, rear hatch and front access panel. The revised front panel featured larger carbon-fibre air ducts that cut turbulence and drag by moving the air around the front wheels more efficiently, while deeper front and rear splitters, and carbon ducts behind the rear wheels augmented downforce by 211lb. Instead of the GT430's extendable rear wing, it relied on just the fixed rear wing, possessing smoother airflow and reduced drag at lower speeds. Top speed was now elevated to 196mph (315kph).

On the basis that 430bhp might be a tad extreme, the GT410 Sport was released to mark Lotus' 70th anniversary. The GT410 Sport emulated the GT430's bodyshell, while incorporating composite front and rear panels but lacking some of the 430s high-downforce sections. Slightly less palatial than the GT430, yet slightly more user-friendly, the GT410 Sport could be ordered as a two-seater or two-plus-two. Power was rated at 410bhp (306kW) and 310lb/ft (420Nm) of torque, enabling 0-to-60mph acceleration of 3.9 seconds for the manual and 4.0 seconds in the six-speed automatic. The GT410 Sport weighed 2859lb (1297kg), making it 62lb (28kg) less than the GT410.

Evora GT

The US market got the final bite of the cherry when, in June 2019 for the 2020 model year, the Evora GT was released. No numbers proclaiming horsepower, just plain old GT. It was only available in North America, and was something of an amalgam of the GT410 and GT430, having a power output of 416bhp

Last Evora on the press fleet, and something of an old friend by close of play 2021, the GT410 pays a visit to Salle Church in north Norfolk. Notice the Yew tree? Évora means Yew tree in Portuguese. (Courtesy Stephanie Ewen)

The final incarnation of the Evora was the 416bhp GT, introduced for the 2020 model year and specc'd for the North American market. (Courtesy Lotus Cars)

(310kW; 422PS) and 317lb/ft (430Nm) torque in manual guise and 332lb/ft (450Nm) as an automatic. The GT incorporated some of the aero sections seen on the GT410 Sport, helping it produce 141lb (64kg) of downforce at maximum speed. The optional Carbon Pack included the carbon-fibre roof panel, rear tailgate/engine lid and front access panel, reducing weight by 49lb (22kg) – quite an advantage, given the heavier all-up weight of 3199lb (1451kg). The optional titanium exhaust brought the weight down by another 22lb (10kg). The manual version of the GT was fitted with a Torsen rear diff as standard, and, as ever, the manual version was 24lb (11kg) lighter than the automatic. Its 0-60mph (97kph) dash was 3.8 seconds in either transmission, with 188mph (303kph) top speed for the manual and 174mph (280kph) for the auto.

We'll see how these cars perform out on the open road in the road trip chapter. Aesthetically, though, did the Evora improve with its successive aero add-ons, carbon panels and gaping air-scoops? Do these venturii, necessary as they doubtless are, contrive to render the frontal image uncoordinated? Or, was the original with its unadorned front and not a duff aspect from any angle a purer expression of the shape? As ever, the original rendition of virtually every model, be it Elise S1 or Boxster 986, is the purest, and truest to the designer's vision.

Cabin of the Evora GT was the usual plush environment, with plenty of Alcantara upholstery and carbon-fibre trim. (Courtesy Lotus Cars)

Pictured at Houghton Hall during the 2011 Hoste Arms supercar outing, featuring luminaries such as Anneka Rice, Mark Blundell and Jack Sears, the original Evora had the purest styling, with little in the way of aerodynamic addenda and ducting. (Courtesy Johnny Tipler)

CHAPTER 6

DYNAMIC QUEST
Ride and handling

Seen here on the Stelvio Pass, the Evora prototypes were camouflaged with black cladding and box-outs to make them unidentifiable. (Courtesy Lotus Cars)

Matt Becker carried out work on the Evora's chassis dynamics and steering, wheel and tyre behaviour, as well as fine-tuning the suspension. (Courtesy Max Earey)

Fine-tuning a sports car's handling is a highly skilled and demanding role, and during the Evora's gestation, that was Matt Becker's job at Lotus. Bearing in mind the company's frequent crossovers into top-line motorsport, there were some exalted predecessors, including John Miles and Colin Chapman himself, who had input into Lotus road cars as well as race car suspension setup.

Driving dynamics exemplifies the car's behaviour on road or track – how controllable it is, how compliant the ride, comfort levels, how it copes with bumps and undulations, what tyres suit its characteristics best, and what are the optimum settings to satisfy a typical owner. In this respect, Matt Becker was responsible for the Vehicle Dynamics test and development work on the Evora project, responsible for the way the car's chassis

The Nürburgring Nordschleife's 12 miles of blacktop with its unmistakable graffiti provided a demanding setting for relentless test laps. (Courtesy Lotus Cars)

system presented itself, through suspension, steering, wheel and tyre behaviour. This is known as chassis dynamics, the action of the suspension on the wheels, tyres and brakes. Another element called chassis statics deals with the chassis support system, the axles and installation cube system. There are numerous new bits of automotive industry jargon to absorb while following the gestation of the Evora, from mule to production car. Essentially, Matt was technically responsible for the test cars, and seeing them through to production.

The key figure responsible for the Evora's braking behaviour, ABS, Traction Control and ESP development was Alan Clarke, Principal Vehicle Dynamics Engineer. Like Matt Becker, he spent countless hours pounding round test sites in the quest for perfect brakes and stability systems to complement the ride and handling behaviour of the Evora. "We didn't just do the ride and handling in isolation," said Matt. "I worked alongside Alan, who carried out the fundamentals on brake and stability system development, and we went along together on joint trips to see how the Bosch systems complimented what I was doing from a suspension tuning point of view, to make sure that all the systems worked harmoniously together." Also typical of Lotus, they worked symbiotically with other project members, particularly the soft-spoken Swede, Ingemar Johansson, who was Executive Engineer in Chassis Design. Ingemar was responsible for the production specification of the parts he had to release, including the chassis systems, brakes, suspension and subframes, and that meant synchronising the work of all the cross-functional teams, not merely to producing prototypes, but with a view to productionising the car. Plus, he was responsible for chassis release, one of the most fundamental elements of the new car.

That transition, from signing off the prototype to

productionising the chassis and componentry of the new car, was a demanding process, because all the systems had to be in place at the end of the prototype stage. Ingemar would, for instance, send a supplier a new shock absorber drawing with the dimensions, and, when available, the setting context. Then, when the prototype was ready to be productionised, the supplier was prepared. Things didn't always work so smoothly though. It was a matter of juggling between the prototype and the final model, so Ingemar might have to ask Matt Becker for the spring's specification, but if it wasn't at the point where it could be given from the test schedule, he might simply have had to make a best guess and use that to enable the next step of the project to move along.

As Matt said, "He chased Alan and me for the specifications on the parts, because he needed to release parts for the next level of build, which was a struggle on Evora with the timescale that we were working to." Multi-talented staff who all knew and understood the basics of each other's business developed a kind of project shorthand, which cut out inefficient and repetitive information flow. CAE was a vital facility as well. Ingemar Johansson explained the process: "When we began the design concept for the suspension, we decided that we wanted double wishbones and, likewise, we wanted very good brakes, and then we said we want fixed calipers – that is another concept – and then when we'd made those selections we went into the dimensions of the chassis and the componentry." Matt took up the story: "Then, basically, we had a huge list of characteristics that we wanted the suspension and braking system to achieve, and Ingemar's team – Mark Crowley, Steve Williams – designed those characteristics into the mule vehicle. That effectively achieved 60-70 per cent of our desired goals. All our initial ideas went into this vehicle, and that gave us a starting point to find out where the problems or issues were." Ingemar continued: "From the mule phase, we went into the EP phase, incorporating the required design changes, and then an EP car was built. That is when it was fully camouflaged, and it remained like that until the car was launched in London." It was a cross-functional procedure: "The design team was made up of 18 to 20 design engineers doing all the chassis bits to get them designed and ready for making parts. The design engineers developed the design engineering solutions, together with the selected suppliers and then they created CAD models and drawings for the release of parts."

As the Evora moved from design to production, testing was ramped up from EP, engineering prototype, to VP, validation prototype. There were three key EP cars – EP2, which was used for establishing vehicle dynamics, EP3, which was the Bosch test vehicle, and EP8, which was the brake development vehicle. The suspension design from the Esprit body mule car was applied to EP2 and measured on the SKCMS rig to ensure that the kinematics and characteristics met requirements. "It was a confirmation that it hit the desired targets," affirmed Matt. Car EP2 was used for steering, tuning ride and handling, which involved springs, dampers and anti-roll bars, Trelleborg suspension bushes, Bilstein dampers, Eibach springs and ARBs, tyre development,

Tried and trusted: like the Evora, the Emira suspension setup consists of upper and lower wishbones, coil-over dampers, and drilled and vented discs. (Courtesy Johnny Tipler)

engine mount tuning, suspension kinematics, and compliance. "Compliance is a function of how stiff the suspension bushes are, how compliant the suspension system is, because obviously if they make bushes that are slightly softer than we want, then the compliance in this system is higher." He recalled that, "The thought process for the Evora's bushes was like the Elise. We wanted to achieve a very high radial rate to achieve high lateral stiffness as you corner the car, but also to achieve very low torsional rates, so like on a racing car, they have spherical joints, which means they have extremely high stiffness, but very low friction and low torsional stiffness. That's effectively what you're trying to achieve with the bush, but obviously you can't have a spherical joint on a road car because all the noise that would transmit through into the cabin is unacceptable. So, we went through probably 10 versions of different bushes, all chasing a very high radial rate, but to get the torsional rate down, and that involved a lot of liaison between Lotus and Trelleborg to achieve."

Then it was handed over to wishbone manufacturers Redaelli and Forgialluminio, who were supplied with the appropriate bushes. "They weren't involved in the design of the bushes," Ingemar pointed out, "because we and Trelleborg did that. We ran a concept design on the bush, and it was decided that you could change parameters by working with the rubber quality only, so that is what we did and therefore they could work in parallel with the wishbone manufacturer."

Turner Prize, anyone? The Evora carried a work of art at each corner in the shape of its wishbones.

These elegant aluminium forgings were made by Forgialluminio in Pedavena, north-east Italy, and then machined by Redaelli near Pescara. They were not merely metal sculptures, but dimensionally specific to the Evora's suspension. In the first instance, Lotus' CAE design process sought optimum stiffness for minimum weight, so the design team came up with a configuration, citing key aspects like where the knuckles picked up and where the bushes were to be located. Targets were defined for the required stiffness, which CAE ran optimisation tests on, enabling them to slowly nibble away at the structure on-screen to identify the best mass solution within the available design space. That produced the basic shape, which was then smoothed off and updated by the design engineers. The finished CAD wishbone pattern was then passed to Forgialluminio, and they created the exquisite reality.

The relationship with key suppliers like Bosch, wishbone manufacturers Redaelli and Forgialluminio, and suspension bush-maker Trelleborg was vital to the Evora's development. "Lotus needed to work with people who were as passionately interested as they were in achieving outstanding results," said Matt. From the outset, Lotus wanted suppliers who were keen to work on the project, who could absorb its concept and could see the mutual benefits of pushing the envelope to get what each party wanted, and all four companies impressed the in-house teams at Hethel with their ability to work alongside Lotus in an atmosphere of mutual respect and learning.

Matt discussed suspension tuning, and the team focus on sustaining that vital Lotus DNA. "In the bigger companies, there isn't usually one person that brings everything together to ensure the car feels complete. Porsche is possibly one exception, where Walter Röhrl signed off everything. Some of the cars you drive feel like they have been developed by many different people, but the car doesn't feel together as one." He did get other people to drive the Evora prototypes, of course, people whose judgment he respected like colleague and racer Gavan Kershaw, whose input he valued as a fresh take on the experience.

The front section of the Evora chassis, consisting of bonded and riveted aluminium extrusions, with wishbones, coil-over dampers, steering rack and brake discs fitted. (Courtesy Lotus Cars)

less familiarised driver could pick up something that had been missed, or even over-emphasised, so that objective outside view provided a kind of ongoing benchmarking," said Matt. The test drivers also needed other people to try out the ergonomics to make sure that the pedal and seating positions were correct, and sufficiently adjustable within the package space. Something like pedal position could not be judged in 20 minutes; it needed at least ten hours' driving to make sure it was acceptable, which is why Matt and his colleagues invited design team members to drive the car on long tests. There was security and satisfaction in signing off a pedal position that had been approved by more than one person.

Long drives are really vital to get the feel of a car, to fully comprehend its foibles – which is why I include some of my own adventures later on. On his first trip to the Nürburgring, Matt made an expletive-ridden phone call back to Hethel to say that the car was simply not saleable with the pedals in their then current position. After half an hour his right leg was in agony because the lateral angle of the pedal was incorrect. Just 10mm or 20mm here or there can make an immense comfort difference to the attitude of a driver's leg. I draw a similar parallel with Porsche's right-hand drive 911s, the air-cooled cars right up to the first of the water-cooled 996s; because of the ingress of the front right wheelarch, the pedal set had to be mounted further into the footwell, so the driver's legs were angled left

As well as a successful racing driver, Gavan Kershaw is Lotus Cars' Executive Director of Sports Car Engineering. (Courtesy Lotus Cars)

Matt was the Walter Röhrl of Lotus – though wryly he commented, "not as well paid." As chief test driver, Röhrl signed off every Porsche prototype before it went into production and, having been up a hill with him at high speed in his black and orange 997 GT3 RS near his Regensburg home in Bavaria, I can vouch for his extraordinary ability. Even when a vast tractor and trailer hove into view, causing Röhrl to get two wheels on the rough, I was more scared for my photographer Antony Fraser than for myself, snapping away as he was, belly-down on the verge. You know you're pretty safe when you're in the hands of someone that's tested the prototype to the point of destruction, and knows to a split-second what its boundaries are.

Same goes for Matt Becker, who played a similar role at Hethel, and was the person in the hot seat on the test track, feeling for the outer limits and tuning the suspension to tease out its most subtle nuances. When he felt the particular direction he was pursuing could be blinkering him, he got another team member – maybe Ingemar, or another chassis designer – to join him on a test run, or what he called an image drive. Then they discussed feel and dynamics to get a more objective view. "The

Veiled prototypes were evaluated at altitude and extremes of temperature, as demonstrated here in Sweden. (Courtesy Lotus Cars)

rather than straight ahead of him. Uncomfortable and somewhat counterintuitive. This was not the case in left-hand drive cars.

"Chassis tuning is like completing a complicated puzzle," averred Matt. "The team had a picture in mind as to how they ultimately wanted the car to feel, and they had to assemble the pieces of the puzzle in the correct position to achieve that result. There was a pyramid of components which was slowly whittled away to remove the characteristics that didn't work and leave alone the ones that did, so that at each stage of development the final car was closer and closer to the team's ideal picture of perfection."

Close of play at EP stage meant suspension hardpoints were confirmed, initial tuning results were to hand, spring rates, ARBs, dampers bushes and steering rack were sufficiently honed to proceed to VP build. All this knowledge was then transferred to the VP cars. As Matt explained, "That was the next level of build; it validated what we had developed on EP, to make sure that when you cross over to the next level of build, everything you have done is working properly, all the correct characteristics you want for VP level are in place. As you got closer to production you went through the different suspension parts that you could change on the car, and then slowly whittle all those different parts out. At the point of production, it was as close as you could get to perfection, within the timescale and the quality of the components."

At various stages of the car's development, the renewal phase and the EP phase, the project moved closer to that ideal. At the end of EP, moving into VP, when all the various departments at Lotus were having their say, the overall result improved the closer it got to production. "Fine-tuning can only be done once the VP stage is completed," Matt pointed out, "because it needed the latest parts, the closest possible to production cars."

Some components were confirmed before that stage; for example, the wishbones and tyres. The ride and handling were 80-per cent complete at the end of VP and before fine-tuning. The final 20-per cent was all about spring stiffness, suspension static geometries, damper tuning, all refinements that could not be completed until the car was effectively a final production model, and this was of necessity a long drawn out process.

During the gestation period there was the task of tweaking the cars between test runs. This was carried out by a dedicated team of Lotus mechanics operating in their own garage workshop. "The technicians in the workshop carried out all the mechanical work on the cars," Matt said, "and they were a fundamental part of the team. There were seven main technicians – James Webdale, Ian Brigham, Dave Jones, Karl Fisher, Daniel Peck, Bernie Warnes and Derek Bradley – and they went with us on all the development trips, so they experienced all of that as well. They operated like the technicians or mechanics: we would come in and say, 'We need to change this, this and this,' and they'd do the work, and we'd go out and drive it again. So, they were the fundamental people who did the hands-on work."

Matt was a keen Nürburgring Nordschleife exponent, and although he claimed that it was difficult to get a clear lap, during development he made it round in an Evora in 8 minutes 25sec, which was directly comparable with a Porsche Cayman S, and just as relevantly, exactly the same time achieved with an Exige S tested by German magazine *Sport Auto* in June 2008. Bear in mind we are talking early days here, pre-production. For further comparisons of contemporary lap times around what F1 Champ Jackie Stewart once called "the green hell," the same magazine's test driver put a Boxster S round the 12.9-mile (20.8km) Nordschleife circuit in 8m 32s in 1999. Robert Nearn did 7m 55s with a Caterham Superlight in 2000, and Walter Röhrl clocked 7m 42s with a Porsche 997 GT3 in 2006. The late Stefan Bellof's 1983 record of 6m 11.13s set in a Porsche 962 Group C race car stood for three-and-a-half decades, till Porsche dropped it to 5m 19.55s in 2021 with a 919 Evo Hybrid Le Mans car. I've driven it a few times, most recently in one of Ron Simons' RSNürburg Cayman 718 GTSs; if you fancy learning the Nordschleife (or Spa-Francorchamps, for that matter), Ron's your man.

Matt Becker's favourite section of the Nordschleife circuit was the Foxhole – Fuchsröhre – and he explained why: "Quite early on during a lap of the circuit, there's a very long straight where you kind of float over a brow, then there is a left and a right, and then you start going downhill, to the Foxhole. It is just incredible because it's flat out, so with the compression at the bottom of the dip you can just imagine all the G-force loads that are going into the car and the suspension. Then you come up the other side, and all of a sudden you have a corner come up on you, very quickly at 140mph. That circuit is just fantastic for showing up any instabilities in your car because the speeds are so high, and you don't recognise the speeds you're doing until you look at the speedo. So it really exposes any aerodynamic or suspension deficiencies within the car. Either you feel very secure and confident – which is a good car, or you feel very insecure and not sure where the car is going – which is a bad car. It's not the best place to do tyre development because you can tell if you like the tyres within the first two miles, but you've got another 12 miles to go to complete the lap before you can change to the next set. But because we were developing the rest of the suspension at the same time as the tyres it was very, very useful being there."

Lotus was also delighted with its Evora testing at Idiada in

DYNAMIC QUEST

Matt Becker helms a disguised Evora prototype around Brunnchen curve on the Nordschleife. (Courtesy Lotus Cars)

JT rounding the Kleine Karussell whilst lapping the Nordschleife in an RSNürburg Cayman 718 GTS. (Courtesy Kostas Sidiras)

Barcelona. Yokohama and Toyota were amongst the investors in a new handling facility at the circuit, which Matt and his colleagues used for handling development. In tyre testing, Lotus used two or three front specifications and two or three rear specifications that could be tested for grip in the dry and the wet. The balance in both conditions needed to be consistent so that the driver was not unwittingly surprised by a change in handling. Idiada had the capacity to provide wet testing on a controlled surface, with a controlled level of water. A purely automotive industry proving ground, racing cars were rarely seen at Idiada, and it had the additional advantages of being in warm and sunny Spain.

Braking points

The brakes on the Evora were another source of justifiable pride for the Lotus team, in terms of consistency, pedal feel and fade resistance. The foundation brake discs and callipers were developed together with AP Racing, and in particular with Technical Manager, Richard Joyce. The Evora's brakes emerged from a concept design process that took into account a number of requirements. First was to size the discs against the vehicle's laden weight, its maximum speed and the package space available, given the wheelarch dimensions, wheel sizes and suspension assembly. There needed to be an accurate match with caliper piston sizes to give good corner braking stability and high-G straight line braking. It was also vital to provide a servo/master cylinder/pedal ratio that produced the desired pedal pressure and characteristics, as well as meeting legal safety requirements. The predicted brake thermal balance and packaging was consistently monitored, and calculations fine-tuned before the components were fitted to the prototype for assessment, iterative development and objective testing.

The Foundation brakes were 350mm diameter discs up front, 32mm thick, and 332mm diameter discs at the rear. It used conventional radial mounts and AP Racing four-piston calipers, and there was also a drum parking brake. The discs were relatively large, but feedback gleaned during testing showed that the brakes were extremely effective, and didn't fade around the Nürburgring Nordschleife. Alan Clarke was responsible for honing brake feel and fine-tuning ABS traction control. During development, he investigated deceleration, assessing the way the car behaved under braking – was it linear, was it progressive, was it too aggressive or not aggressive enough? He worked on balance – did the car brake better turning into in a corner, or was it better to do all the braking in a straight line? It was a constant process of assessment and adjustment. In terms of circuit driving, he was testing for overall performance in fade and resistance, and on the Nürburgring where there are so many undulations, curves and adverse cambers, he needed to assess the ABS, whether it helped keep the car in the corner, and whether it oversteered or understeered under braking. The Evora had radial-mounted calipers, allied to stiff uprights, taking out another opportunity for flexing and movement from the suspension. The result meant responsiveness was improved, and the driver could feel that the car was behaving better. Alan went through a checklist, ticking boxes to achieve the braking goals. "I verified brake balance, and tuned brake pedal feel," he said, "which was dependent on servo characteristics, pedal ratio and brake pad selection; thermal performance including fade resistance, driver feedback and confidence, noise and modulation to give it the Lotus DNA." All of this was carried out during countless laps of Hethel test track, Idiada, the Stelvio Pass in the Alps, and three visits to the Nürburgring. Alan flagged up that the car's brakes were sized to ensure that they have no fade problems and excellent stopping performance. He testified that they measured up to 1.3 G deceleration from V-max, a class leading result, and affirmed that Bosch were the best ABS and traction control specialists he'd ever worked with, because of their openness and willingness to experiment to achieve the brake performance Lotus sought. During testing at the Nürburgring, Alan worked with the Bosch engineers to adjust the ABS characteristics and amend the thresholds to alter the amount of pressure being applied to the rear brakes to stabilise or destabilise the car, as required by the conditions. There were more than 10,000 different parameters that could be changed, just within the ABS system, and then there was traction control. On his Nürburgring trips, Alan clocked up a considerable number of laps, as well as the journeys there and back on the same set of brakes.

Arctic Circle

Most manufacturers seek a modicum of privacy when developing a new model, and the more remote and inhospitable the better. A raw location also has the advantage of providing an extreme environment in which to hone handling and drivetrain that would be impossible in normal driving conditions. An unpronounceable name is even better: it adds mystique and inhibits snoopers. Welcome to Arjeplog in northern Sweden, a stone's throw from the Arctic Circle, where an Evora prototype was fine-tuned at Bosch Engineering's Vaitoudden Winter Test Facility, a kilometre or two outside Arjeplog.

While conditions are relatively consistent at Arjeplog, in so far as it's sharp at -7°C when I visit, and snow is a given in winter, there's still variety on these remote Lapland roads where slush, ice and bare asphalt present different friction levels. Other, less predictable, natural hazards lurk in the forests too. As Bosch's test driver and project leader Alex Böss put it, "The car has to be stable when you're braking, particularly when an elk or reindeer steps out in front of you."

Bosch may also be synonymous with dishwashers and power tools, headlamps and wiper blades, too. But its Engineering services subsidiary has long led the way in automotive engine management hard- and software, ABS and traction control systems. It's 50 years since it launched the first ABS (anti-lock braking system) in the Mercedes-Benz S class. And that's how long the German manufacturer has operated beside this frozen lake. A decade later, its traction control system was up

Bosch's test driver and project leader Alex Böss: tirelessly dancing on ice. (Courtesy Jason Parnell)

Bosch developed software for the Electronic Stability Programme system on the Evora mule, including basic ABS and TCS. (Courtesy Jason Parnell)

and running, and in 1995 it launched its Electronic Stability Programme (ESP). When the Evora was in development, around 250 technicians worked in the laboratories and offices at Arjeplog, and though the majority were from Germany, there were a few representatives from the global automotive industry present, including Lotus. It depended whose cars they were working on, and there was a pretty good cross-section there on my visit. Bosch staff's Arctic role was transitory, averaging 12 weeks over six fortnightly stints, swapped for the less austere climes of their Stuttgart HQ.

This was the first joint venture for Bosch and Lotus, beginning in summer 2007, when the Eagle project took off. Bosch was tasked with developing and implementing the ESP system on the car, bearing in mind US legislation, which included a basic ABS and TCS (traction control system). The Evora spec was very different to the Elise and Exige, so they started with a clean sheet. Though engineering principals were similar, there were no shared components, so everything was brand new as far as the Evora's systems were concerned. The way it worked was, Bosch Engineering developed new functions for one manufacturer, and installed the same hardware in different clients' cars, then fine-tuned it according to specific requirements and applications. That made it vastly cheaper for the recipient, as there was no need to develop new hardware, electronics and hydraulics, simply to adapt and develop the requisite software for the fresh installation. It wasn't that Lotus lacked the expertise to create and implement such componentry itself, far from it: the company pioneered computerised suspension in the 1980s and the Lotus traction control system (LTC) entered production in 2005; the chief obstacle was that it was just too expensive to

Ski-bobs provided an entertaining diversion whilst visiting Bosch at Arjeplog. (Courtesy Jason Parnell)

manufacture the hardware from scratch for the new model, hence the involvement of Bosch. "We would adjust and apply an existing system to a new car," said Alex. "On the one hand it was cheaper, and on the other you had proven systems to work with on a totally new car."

The mule was supplied by Lotus, a well-used Esprit shell, cut-and-shut to fit, tantalisingly cladding the Evora's chassis and driveline, though the engine note was a distinctive six-cylinder roar, unfamiliar in what, to the casual observer, looked to all intents and purposes to be an Esprit. After 2007's technical discussions and negotiations on electronics, the two companies agreed on the most appropriate systems on which to base the Evora installations. "Originally, we had a lot of issues to discuss for the electrics," explained Alex, "because Lotus never had a big CAN (control area network) protocol for the different parts working together, so we had to find out which signal worked best for what application, in order to establish a network for the ECUs in the car so that they could 'talk' to the engine and the dashboard." Bosch's first act was to implement all the measuring equipment and establish the engine management communication, and the mule was taken to Spain to undergo the high friction phase on the Idiada test track. "We generally tested cars there during the autumn and winter season because it's quite dry. So, the Evora project started not in Sweden but in Spain."

Ironically, fickle weather meant a late start at Arjeplog, as the lake was only accessible from late December. "The temperature was cold enough, but there was too much snow," recalled Alex. "When the snow covers the ice it will insulate it, so the crust won't get thicker, and it's a Catch 22, because you can't go on the ice and clean the snow off as it's too thin. So, actually, you need cold temperatures without snow at the beginning, and then later on, some snow is nice." By December, it had frozen to a depth of 50cm, thick enough to support even the trucks and buses for which Bosch was producing ABS systems. Shades of *Ice Road Truckers* on TV. "We test the cars on the same tracks as the commercials, and when you stand next to the car you can feel when a truck is coming because the ice goes up and down.

Opposite, clockwise from top:
Interminable snowscape: a boat and signpost provide scant relief in Arjeplog's frozen wilderness. (Courtesy Jason Parnell)

Whilst most of Arjeplog's frozen lake has a covering of snow, Bosch keeps a central circuit snow-free so test drivers can evaluate traction control on bare ice. (Courtesy Jason Parnell)

Dog-sleds are a normal form of cross-country transport in Swedish Lapland. (Courtesy Jason Parnell)

It's like the earth's mantle, which is made up of plates, and the water currents underneath the plates move them on the lake surface, making little bumps."

At lunchtime, the Bosch crews went off to an octagonal wooden café beside another nearby frozen lake, where meatballs, chips and other sturdy Swedish morsels were barbecued over a roaring central hearth. Back at the Bosch Centre, it was time for an outing. The prototype's cockpit was akin to a racing car, and nothing at all like the Esprit that it purported to be. I was the passenger: negotiating the roll cage and bucket seat was the work of a gymnast, and I endeavoured not to nudge the open laptop mounted ahead of me that logged the telemetry. The gear selection linkage was exposed and mounted higher than it would be in an Esprit, while most of the instrumentation was a typical prototype lash-up, with lots of cables bundled chaotically underneath. As soon as the car was in motion, the laptop screen came alive, displaying a multitude of different coloured lines that registered the slightest twitch. "Traction control is always working out here," Alex said, which was reassuring, though not entirely convincing, as at the time he said that, I was looking out of my side window in the direction we were travelling. "All the time we're testing it we are logging data, seeing what the effect is and making minor adjustments. We take a measurement, and we'll have a subjective feeling about it and if it's not good, we'll stop and assess the measurement, interpret the details and change the parameters in the computer and download them on the ECU."

There was a marquee out on the ice, looking like something out of a village fête; the only difference was that it was pitched in the middle of a frozen lake rather than a meadow. God help us if they use tent pegs, I thought, though presumably they'll be frozen into the ice. "This is the presentation area," Alex told me. "It's where we present the cars to the manufacturers when it's handover time." Alex drove confidently on the white stuff – as you'd expect. I glanced across at the speedo. We were doing 100kph, sideways. It was a left-hooker, so I assumed we were in kilometres rather than miles-per-hour, but it was hard to be sure in this unfamiliar all-white context. Like the snow in these parts, the thrills came thick and fast. There were no regulations on the number of cars that could be on the lake at any one time, but it did depend on the thickness of the ice. Cars had to be parked at least 5 metres apart so that the weight wouldn't be too concentrated. "Maximum speed is meant to be 70kph, but that is slow," said Alex. "We need it to be more, as nobody is driving at 70. Normally you can drive here at 90."

There were two handling courses out on the ice, each nearly 2km, plus a dynamic area, which included firmly packed snow, rough ice and black ice, and the vista was flat as far as the eye could see. The circuits were delineated by snow banks, which we didn't want to hit, as it would be a bore having to wait for the on-site Land Rover to come and haul us off. He deftly slid the Esprit-Evora onto the snow circle, a 300-metre oval with a glistening inner ring, polished by tyres, abutting the outer compacted snow circuit. In no time at all we'd done two 360-degree spins, but mostly we were broadside one way or another as he went from lock to lock. The front wheels caught the ice rink and we pirouetted. Again, and again. Maybe he's become snow-blind, I wondered. But no, he was clearly loving his job. When testing the car for real, as opposed to looking dramatic for the benefit of Jason Parnell's lens, the telemetry told him what was going on and he responded accordingly.

According to Alex, the Lotus was a relatively easy car to drive, even at Eagle prototype stage. What constituted a difficult one, then? A very heavy one, like the Porsche Panamera or the Bentley Continental, both in evidence at Arjeplog. "They are very powerful, but when you drive them on the ice, for example, and something goes wrong, it's too late, you have no chance to get the car back on track." I noticed that certain sections of the Esprit body were slightly the worse for wear, though it was merely the mule. Same for one of the Bentleys back in the car park. "The Lotus only crashed once this season, but nothing was broken because the snow is normally soft, and if you're lucky it's just bouncing off the snow bank and you're back on the track. Yesterday, there were very bad conditions and the tracks were closed because of a heavy storm, and we couldn't see anything on the lake, and also we had snow banked across the track because of the wind."

Back on dry land – it's the knolls and trees that give it away – Alex brought the Evora mule to the site's test hills. Less extreme that Chobham, say, there were three differing gradients, 10 per cent, 15 per cent and 20 per cent, covered over by open-ended barns. Up the left-hand side of each run was a 25-metre strip of solid ice. He positioned the car at the start of the hill, but it slid inexorably backwards when he attempted to pull away. There was no grip whatsoever. Then the traction control system came into play to stop the inside rear wheel spinning and took the power to the outside wheel, which was on asphalt to the right-hand side where it sensed it would get some purchase. Then, with a bit of roar, we rocketed up the hill. It was a simple but effective demonstration of the car's prowess. Alex quoted another example: "Quite often up here (in Lapland), on normal straight roads, on one side you have asphalt and on the other you have ice or snow where you have no traction, and that shows how you also have this kind of control. In a sharp corner,

facility. We have to drive there because we want constant conditions, so if we have a problem with the software on Monday, we want to re-test or verify it on Wednesday in exactly the same conditions. So, to make the fine-tuning and the last calibrations on the system we need longer roads to get a good feeling for the car. You need time for that." And there is a time limit, because by April the ice will have melted. Normally Easter is the cut-off point when work stops at Arjeplog, and technicians head back to Bosch sites in Germany.

Traction control
When driving fast down an icy road, dynamic stability control rather than traction control comes into play. The car starts to spin, and the dynamic stability control applies inner and outer brakes to get the car to the correct attitude as instructed by the steering input. Without this facility the driver would need four brake pedals to get the car orientated in the correct manner. In the 100mm diameter circle on the Swedish ice, the inside was more slippery than the outside. So, if you drove on the outside at 55mph, you'd oversteer, but if the driver kept his or her foot to the floor, and held it there, the car would veer onto the polished glass on the inside. The hill-climb in Sweden was on the road, about two miles long through woods, with a very steep downward slope at the end. One for the brave. The traction control system developed between Lotus and Bosch engineers for the Evora is described as integrated dynamics. Every modern car has intervention systems, but on the Evora the ABS and traction control systems have been developed to the point where small variations can be perceived and corrected. The system

Scandi-Noir: on the way to Arjeplog I cruised this Exige around Stockholm's inner archipelago. (Courtesy Jason Parnell)

normally the inside wheel will spin so you lose traction, and it's the same there: you are going to brake the inside wheel and transfer the torque to the outside wheel."

All the work at Arjeplog is carried out in the garages and offices, where there's support from the measurement teams when they hit problems with equipment, or the car is broken. "We'll do the fine-tuning on the curves and on the open roads

With a background in motorsport, Gavan Kershaw is Lotus' Executive Director of Sports Car Engineering, and oversaw much of the development of the Evija and Emira. (Courtesy Lotus Cars)

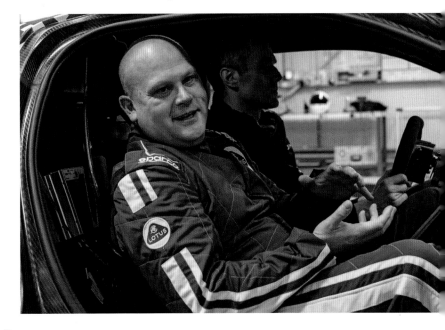

recognises understeer, so that if the driver goes too fast into a roundabout with too much understeer on, the car will react and subtly reduce engine torque on the driver's behalf. It's all part and parcel of what makes the Evora such a superbly handling car.

I've chatted at length with Gavan Kershaw, who is currently Lotus' Executive Director of Sports Car Engineering, and he declared, slightly ruefully, I thought, "The Evora was probably the last of the cars that got developed without all of the software, all of the active suspension, rear steers and things like this. We were one of the few manufacturers that allowed our cars – the Elise, the Exige and the Evora – to be turned off completely. We kept full ABS, but the ESP system was completely turned off, and it was then down to the driver. The cars that are coming towards us now have four electric motors, so you can physically have as much torque going independently to each wheel." Gavan

Completed cars approach Lotus Buy-Off for finishing touches, such as – in this case – the badge to be popped on. (Courtesy Lotus Cars)

and Matt Becker joined Lotus on the same day, and as Director of Attributes and Product Interior, Gavan now occupies a similar role to Matt's of a decade earlier. "It's about the way the cars drive and the way they feel, and that's all signed off by me," said Gavan.

Neil Thomas interview
Evora Platform Manager Neil Thomas joined Lotus Cars in December 2011. Prior to joining Lotus Engineering, he was at Land Rover, in Concept Engineering as Launch Manager for Freelander and Evoque. "We were doing lightweight concept vehicles based on the same technology as the Elise, so the chassis that we designed and put under the vehicle was largely based on a bonded, extruded chassis frame with composite body panels, a 1250kg lightweight Defender, and examples of it now reside in the museum at Gaydon. So, I'd already been indoctrinated in lightweight thinking before coming to Lotus. The reason I came here was on the back of the Dany Bahar era. The line-up of cars he was proposing to build was a little bit ambitious, but at least if one of them came through, it seemed like it would be a great place to go: the Esprit in particular was a beautiful concept.

"I began in the engineering division, working on the Hydrogen Taxi which we re-engineered and manufactured five vehicles for the London Olympics in 2012. Next, I joined Richard Rackham as the Platform Manager and then Launch Manager to put the Evora 400 into production. Even during that period, the instability was apparent; especially the lack of investment, but that's definitely different now. It was a transition period, so Dany Bahar was in his descent, and the new ownership for Proton and DRB HiCom, then consolidation through the Jean-Marc Gales period, and then on to the present Geely years. We're already starting to see the fruits of that, and at least one new model was conceived at the onset of the Geely ownership. That would never have happened without the investment and stability to enable it, so in the next year or so you'll start to see the replacements for the current cars beginning to be developed. We've a succession plan, and we're looking at a five- to ten-year horizon, rather than the previous two years."

Hethel – Lotus' test track
One of the principal reasons why founder Colin Chapman chose Hethel as the location for Lotus Cars and Team Lotus back in 1965 was the disused airfield. Roaming the skies over Norfolk in 1966, he spotted the site and immediately the plot was hatched

The Type 49 in its original unadorned colours, and with zero aerodynamic aids, gets a run out on the Hethel test track, in the hands of Classic Team Lotus' Team Manager, Chris Dinnage. (Courtesy Johnny Tipler)

to relocate from Cheshunt's north London suburbs. Not only did he recognise Hethel's potential as test track, he visualised a future when visitors would be just as likely to arrive by plane as by road. That, allegedly, is why the reception doors face the track rather than the road.

Hethel airfield was built by the American Air Force in 1942 on the site of Hethel Hall. Arriving in 1943, the USAF 389th Bombardment Group flew B-24 Liberators as part of the Eighth Air Force's strategic bombing campaign. The former aerodrome's control tower is now the Clubhouse, overlooking the 2.25-mile circuit, and the layout still corresponds – with adaptations – to the plan of the wartime airfield. Part of the main straight is the original main runway. Over the years it's hosted countless Team Lotus testing sessions with heroes like Jim Clark, Jochen Rindt, Ronnie Peterson and Mario Andretti giving it their all. Successive Lotus road cars have been fine-tuned here, no more so than the Evora.

The track was redesigned in 1999 by test driver Alastair McQueen, incorporating a series of bends known as Clark, Hill, McQueen and Senna, with a turning circle for hectic steering and balancing practice, while the Main and Factory Straights and the North Hairpin are ideal for higher speeds and coned slaloms. By 2012, it was radically reconfigured, with many attributes up to FIA standard, and employed during the following decade as a proper facility for honing and fine-tuning production cars, as well as hosting Festivals and demonstrations, including the 2021 unveiling of the Emira.

The Lotus Driving Academy
To get the most out of their Lotus, owners and potential customers can familiarise themselves through sessions in the Driving Academy, held on the Hethel test track. There are four levels to the Driving Academy: Bronze, Silver, Gold and Platinum. Bronze is a good introduction to safe road driving as well as track driving. One of the hosts is Lotus aficionado, commentator and specialist Guy Munday. "We use the North Circuit for some of the morning sessions, and the South Circuit for some of the morning sessions, and then in the afternoon we open the whole track up for the complete 2.2-mile experience," explains Guy. "They're 20-minute sessions, so customers get three 20-minute sessions

Lotus Driving Academy gives potential owners and marque fans the opportunity to hone their skills under supervision on the test track. (Courtesy Lotus Cars)

in the morning and two 20-minute sessions in the afternoon. We divide the group in two, so half are driving, while the other half come with me on the Factory Tour and Classic Team Lotus tour. Then, halfway through the morning we turn the group around and go again! The number of laps they get depends a little bit on the ability of the driver, on how quick the drivers are. They're timed sessions rather than distance covered, so when time's up, all the cars come in regardless of the number of laps they've completed. It's non-competitive, so we try to space the drivers out around the circuit so they shouldn't really encounter other traffic. It's pretty much teaching you to be a safe road driver rather than an out-and-out race driver."

Lotus buff Guy Munday drives the Evora Safety Car at 2010's Classic Team Lotus Festival at Snetterton. (Courtesy Johnny Tipler)

The Academy instructs on how to make use of oversteer and understeer. "The third discipline utilises the low-grip circle, and on a Silver day we encourage understeer and oversteer, to show you how to get out of it once it's happening. Those three disciplines become relevant in the afternoon, because we open up the full 2.2-mile track, and you put into practice what you've learnt in the morning. For instance, at the end of the straights you will be braking – and braking as late as possible and as straight as possible, which you've learned on the brake and steering discipline in the morning. The complex of curves shows you how to look through the first curve into the second and third ones to get your positioning and speed right so you don't lose momentum. There's a hairpin at each end where you can put your understeer and oversteer into practice, so it's all relevant to safe road driving, but also track driving to a degree as well. Also, we teach heel-and-toe; everybody has heard about the technique, but very few people really understand what it is and what benefits it brings, so we teach them how to do throttle-assisted gear changes, how to execute later-braking into the corner whilst taking the down-change in a nice smooth manner. The afternoon track time brings slightly more advanced techniques into play."

The culmination of the Driving Academy is the 'Licence Weekend.' If you successfully complete the Bronze, Silver and Gold, then you're awarded your Driving Academy licence. "If you do all those days, plus the Platinum, then we call that 'The Carbon Weekend,' which is the ultimate experience currently on offer. We use our cars on the Lotus test track so potential customers can see whether it's exactly what they want, and from Lotus' point of view it's a good incentivising conversion tool as well. We've had the Evora, and four Elises, and there's also an Exige 410 which we use on the Gold and Platinum days." There are a dozen instructors. "Most of them are active racing drivers, and we use Lotus' own staff as well, and you get a slightly different experience from them. There's a difference between what a racing driver can teach you and what somebody like Dave Minter who's been Lotus' ride and handling guru for 40 years will teach you. You get a different experience from them than you do with the race drivers: they tend to be more sympathetic on the cars and encourage a smoother drive. Dave and his revered ride-and-handling colleague Alastair McQueen set up the Driving Academy back in 1998, and it's run pretty well. Alastair was very much responsible for the basic format to which we still adhere to this day. The emphasis is very much on safe road driving, and it's worth pointing out that the Driving Academy is responsible for selling a good number of cars to punters."

Above (left and right): Lotus Cars marketing executive, Kristie Becker, was production coordinator of a documentary on the Evora's genesis, including filmed footage of the prototypes in testing on the serpentine Stelvio Pass, and the Evora launch at ExCel. (Courtesy Kristie Becker)

Posed with the first customer Evora sold are three of the creators of the DVD recording the car's genesis, Lotus Marketing Executive Kristie Becker, who was production coordinator, David Bainbridge (right), director and producer, and Stuart Nicholls (left), associate director, camera and editor. (Courtesy Kristie Becker)

Matt Becker honed the Evora's ride and handling characteristics to perfection. (Courtesy Max Earey)

Matt Becker – Executive Engineer, Vehicle Dynamics

Two strong hands on the steering wheel: not a member of the Dixie Chicks, rather Matt Becker, Lotus' Executive Vehicle Dynamics Engineer, who achieved the Evora's impeccable ride and handling. Till he left the company in 2014 to perform similar duties for Aston Martin and thence, in 2022, to Jaguar-Land Rover, Matt was a Lotus man through and through. Son of Roger Becker and brother of Kristie, his career began at Lotus in 1988 as an engineering apprentice learning key areas of automotive engineering. Initially, Matthew spent his early career working in the powertrain engineering consultancy division, building and developing engines.

In 1995, he became a junior engineer in Vehicle Development, joining the Elise development team alongside Dave Minter. Matt was a key figure in the ongoing improvement of the Elise; he worked on the Metal Matrix Composite brake discs and on vehicle performance testing for the Elise Mk 1, and collaborated with Yokohama on the A038R road and track tyre that was first produced for the Elise. Matt continued developing the ride and handling on some of the most memorable and critically acclaimed Lotuses, including the Esprit Sport 350 and many Elise variants, including the Elise Sport 135, Sport 190, Elise S2, 340R and the 2-Eleven. He also played an integral part in the press launches of all the cars he was involved with, and he continues to be well respected within the world of motoring journalism; I was on a Porsche shoot in Snowdonia with Antony Fraser on the 'Evo Triangle' and we were repeatedly buzzed by an assortment of high-end SUVs. Breaking for lunch at the Llyn Brenig café, there were all these SUVs in the car park, and inside was Matt Becker, by now working at Aston Martin. He and his crew were evaluating rival SUVs ahead of Aston making its own DBX.

Matt had overall responsibility for the vehicle dynamic development of the Evora project: a task that included ride and handling, and sign-off for the ABS and ESP characteristics, ensuring the car has a holistic feel in all dynamic attributes. I spent months watching the Evora prototypes being refined in the factory, and my first ride was with Matt. We clocked up a few rural Norfolk miles for the benefit of Jason Parnell's car-to-car photography, and headed back to the factory test track. With press day still several months away, I could hardly believe my luck as I eased myself into the body-hugging Recaro seat and buckled up. This was going to be a wild ride.

I eased the black beauty out onto the Hethel apron. The test track is a mixture of sinuous and straight, and the twisty sections are no wider than a small circuit like Abbeville or Cadwell Park, while the main straight is part of the former aerodrome's main runway, and by far the widest part of the track. At the top end is a roundabout for pirouetting, at the other a sweeping one-eighty for full-blooded powerslides. Speed grew, lap by lap. No slowing down for the tyre chicane two thirds the way down the long back straight, deftly opposite locking sideways in the horseshoe corner, race car-style late braking into the complex, jinking this way and that, clipping apexes, sometimes a wheel deliberately over the edge, but always firmly in control; hard on the anchors into the outer half of the roundabout, flowing into more tail-out and opposite lock, and then flat through the kinks leading onto the back straight again. Handling so flat going into and out of corners, unswervable under braking and responsive to hard acceleration, it had everything a sporting driver could wish for. Traction control on or off, it responded predictably either way. Spellbinding!

Time for a bit of analysis. The Evora VP prototype I took for a spin at the Hethel test track was already endowed with correct suspension hard points, production-intent bushes, production-intent springs, and its anti-roll bars and damping characteristics were almost refined. The sports suspension was still being tuned to blend the suspension to the weight and weight distribution of the production model, so Lotus was giving the car its final polish for production specification. As the Evora entered production, sample production cars and sample components were extracted off the line, and checked and measured against those standards set by the VP development prototypes. As Matt

Op-art patterns disguise an Evora 400 on test at Hethel's backyard track. (Courtesy Lotus Cars)

described, "With production tolerances, you are always going to get some differences in certain aspects of the suspension or the components. So, you drive and evaluate the production cars to make they are within an acceptable tolerance with what we wanted to achieve for the car's dynamic behaviour." It's an ongoing process.

Nevertheless, it's fair to say that Matt's expertise in tuning suspension was stretched by the demands of the Evora. With the two suspension settings, base and sport, his intention was to maintain some continuity in feel across both. He compared the result to the Elise and the Exige; its base-level suspension was similar, but slightly more comfortable than the Elise. Sport suspension was more like the Exige, but with an increased level of response with the steering and improved body control.

Interestingly, Matt also explained how the longer wheelbase V6 Exige came about: "Some of the original mule cars for the Evora were actually stretched Exiges. I think that was my father's (Roger's) idea. He was in the workshop one day, and he looked at the Exige-based Evora mule cars, and he went, 'Why don't we do that? Why don't we turn the Exige into a V6, just like that?' And so, the original sketches for the Exige V6 were actually based on the Evora mule cars. And then, it turned into something quite different. He was like, 'Well, actually, the engine's not that expensive, and it delivers the performance we want.' Yes, there's a weight penalty, and we were concerned about the balance of the car, because the weight is actually quite a long way to the rear, based on that wheelbase. But that's where the concept came from."

Much of Matt's time was spent at the Nürburgring, working on the Evora's tyre development. As ever, it was that repetitive process of testing, discussing, changing and testing again which brought the eventual result that Lotus wanted. The test car was shod with specially-made Evora-designated tyres, which Matt was using to finalise the suspension damper settings, so he was trying to be kind to them, drifting notwithstanding, because the rest of the system had to blend to those tyres. At this stage of the VP process, he was fine polishing it to come up with the production settings, so he would continue to use one set of tyres until he could put the other set on to confirm that tread depth did not make an excessive difference to response. "As a tyre reduces in tread level, the response it provides rises," he says, "because the block stiffness goes up as it gets closer to the base of the tyre." As things turned out, the host of experimental tyres the team worked with were shelved, and the Evora hit the streets shod with off-the-shelf Pirelli P-Zeros.

Tyre selection
Here's how that happened, in Matt's own words: "In 2007, we created a mule car, which was based on an aborted Esprit project, which was basically a cut-and-shut job with an extendable wheel base. Underneath that was an Evora layout, so the wheelbase was correct. The first cars didn't have ESP (electronic stability programme). The first cars had ABS and traction control, but they didn't have full ESP. Because of the timing of the project, we didn't have enough time for there to be a full ESP system. When we went to the Federal car, we had to go full ESP. That's why there was a delay to launch the Federal cars. We started on tyre development, and we decided on a 215/255 tyre balance with Yokohama.

"Quite quickly, we realized the front tyre, whatever they did with it, just didn't have the capacity we needed. It didn't have the balance; it had too much understeer. And they had made a specific tyre. We decided we needed to go for a wider section of tyre here, which in simplistic terms, you'd think, well, that's okay. But actually, it's not, because that has an effect on the rest of the car, because the design team have designed the car around a 215 front, so suddenly they've been given a wider tyre which they have to make look pretty within the wheelarch. Yokohama told us, 'You're just not going to get the capacity of the front tyre that you need to achieve the balance you want. We've thrown the highest compound we can, as much cornering performance as that front tyre can deliver.' Then they said, 'Okay, we need to go to a 225/40/18 tyre then to give you the capacity and the balance you need within the car.'

"But, anyway, back to the tyre sizes. We ended up with a 225 front, 255 rear. I must've spent two years developing the tyres with Yokohama, but never got what I wanted. Because I always had this problem: I couldn't get the rear of the car to feel hooked up. You apply the steering, but then you always had this two-stage response, which I hate. We'd done all this development with Yokohama, and I don't know how many different specifications of tyres we went through. I was at the end of my tether; I just couldn't get the car to feel the way I want it to feel. And then, a Pirelli guy that I knew said, 'Why don't I send you some off-the-shelf Pirellis? Same size tyres, 225, 255.' And he sent me these Pirelli P-Zero tyres. So, I said, 'Let's just give them a go.' We bolted them on, and it was like night and day! Suddenly, we had these tyres that actually made the car feel agile. I was just like, 'Well, Christmas has come early!' The problem was that Yokohama had already made about 400 tyre sets, and I've just bolted these Pirellis on. And the car was transformed. One of the key things I was tasked with was to achieve incredible dynamics. Which we did. But first I had this awful conversation with my father, because he (Roger Becker) was the Technical Director for all of engineering, for Esprit as well as Evora. I said, 'Dad, I need you to drive the car with the Yokohamas on and the Pirellis on. You can do it blind, if you want.' He said, 'I don't need to do it blind. Just give me the car, and I'll drive it on both sets.' And he came back in, and he went, 'It's night and day!' He said, 'We have to go with Pirelli.' And, of course, they're rubbing their hands, because they didn't have to do any work whatsoever. Straight off the shelf, they were. I think it was actually a Mercedes specification tyre. So, it was a bespoke tyre, but for Mercedes. And it just worked on our car. It's not easy just to switch tyre suppliers, because all the legal certification and everything was done with Yokohama. But suddenly, we had to switch to Pirelli because they gave us so much better performance."

In such an aggressive programme, time-wise, Matt worked closely to develop and blend the suspension system with his colleagues, and he was very appreciative of the work of the whole team in bringing the Evora to fruition. The structure, the bush system, the tub, the brakes, the whole package requires a tight integration of team activity, and he singled out Ingemar Johansson in particular for his input in designing the suspension system. Even after he'd moved on to Aston Martin, Matt could eulogise about the Evora, and I quote, "The Evora project for me was the pinnacle of my career, not in technical content as cars have progressed so much since then, but it was a project where at that point I took all of my experience, and as part of the team, we developed something really special."

Back on the Hethel track, Matt introduced me to the Evora's traction control facilities. We circulated with increasing drama, passing from passive mode, which had everything buttoned down, into sporting mode where all hell could break loose, were it not for his spontaneous reactions keeping it under control. In fact, the TC system had two modes. Standard is where the traction thresholds were set to be relatively safe, so it detected any wheel slip and reduced the engine torque. Standard mode also had understeer recognition; if you were approaching a roundabout, and during turn-in, the car started to understeer, the system recognised that the deceleration was not increasing but that your steering angle was increasing. So, it then reduced the engine torque to stabilise the car. In sport mode, traction control turned off understeer recognition and raised the traction thresholds to allow the driver more autonomy; the system remained in the background to provide a certain amount of control, but the slip thresholds were raised to Sport mode to permit more yaw angle, more slide. The ABS kept on functioning, and remained at a constant level throughout all traction control modes. As in the Elise, the Evora's ABS system was configured so it only ever intervened when absolutely necessary, and the Evora certainly engendered an aura of invincibility when travelling quickly in standard mode.

The Evora was 450-500kg heavier than the Elise, but the challenge for Lotus was to make it feel as light and agile as its smaller sibling. Thus, it manifested a sense of quick direction change and agility, which would deny its weight if you weren't already familiar with an Elise. The sports suspension brought the turn-in quickly, the car maintained a sense of linearity in the steering, while an over-aggressive steering response made it difficult to drive with precision. It was a delight to drive, skipping, almost floating imperceptibly through the Hethel chicanes. The Evora wanted to be adjustable on the throttle, to balance the attitude of the car in the corner, like you could with the Elise. But it was also important that it didn't over-react as it responds to the throttle. At 100mph it was very comfortable, relaxed and informative. With foot flat to the floor, as understeer increased, the engine torque was reduced to stabilise the car. The throttle on-off requirement was to make sure that, if you happened to be in the middle of a corner, you didn't accelerate dramatically and dangerously, which was another carefully considered safety detail. With a lovely feel to the gearshift, the Evora came across like a bigger Elise; instant throttle response, invitingly chuckable, but more absolute grunt since the engine had twice the capacity. As far as the driving experience went, the Evora was a larger, more dominating version of its little sister. There are crucial differences, naturally. At 5-6000rpm the Toyota V6 was really roaring, blaring a glorious six-cylinder engine note, with the extra 400rpm unleashed by the Sports mode.

Matt's job at Lotus was one that many would envy, testing fabulous cars day in, day out, but it demanded a very particular kind of dual attention. He had to be able to drive the car to its limit with one part of himself, while another was detached enough to analyse its performance.

Matt questioned Hethel's inherent processes, however. "If I'd have stayed at Lotus, I'd have less than half the experience I've had since moving to another company. Other manufacturers have all these processes and procedures in place, which wasn't the case at Lotus. But maybe you do need them, because if you want to create a car as robust as a Porsche, you do actually need to have processes and procedures, and delivering methodologies to basically deliver cars that can do what, say, a Porsche can do, in terms of quality, reliability, and performance ability. And at Lotus, we never did the studying we needed to do, probably because we didn't have the money to be able to create the cars that customers in volumes really wanted. The Elise gets away with it, because the Elise is a special car. It's got a focus of what it needs to achieve, and I think the Evora never had the attention it needed in specific areas to make it a success. It sounds negative, but it's not. That's the reality of what happened behind that car. I think with the Emira they learned their lessons." But the bottom line was that, of course, eventually, he did sign off the Evora.

Tempting as it was to stage a launch at Évora, the city that shared the model's name, in Portugal's rural Alentejo region, or even the major centres of Lisbon or Porto, the economic downturn in late 2008 decreed a UK launch. Notwithstanding international connotations deriving from the car's name, the car was launched in London at the 2008 Motor Show, and Scotland was the destination for the first drives as part of a two week launch event. Matt Becker described the reasoning behind a Scottish launch as opposed to, say, the Algarve: "The reason we chose Scotland for the launch was because the roads are quite bad, and actually, one of the reasons we chose it was deliberately not to use a smooth road or race circuit in Europe. We thought, 'Let's go somewhere within the UK where we can actually demonstrate the ride quality of the car and the ability of the car on UK roads.' The biggest problem was it rained. We had sideways rain for two-and-a-half, three weeks. As long as we were there, it just rained and rained. But I think all the journalists got that the reason we used that area was because there are some fantastic roads, some fantastic scenery. Like, Glencoe

Glamour girls grace the Evora stand at its unveiling at London's ExCel in 2008. (Courtesy Lotus Cars)

Policemen interrogate a faceless person at the Evora launch at the ExCel Centre, 28th July-3rd August 2008 – a gimmick to stir interest. (Courtesy Lotus Cars)

An Evora snapped during a media event in Provence. (Courtesy Antony Fraser)

and Rannoch Moor, there's some fantastic roads up there, and that challenged the car. We wanted to demonstrate how good the ride quality was and how isolated we were, to demonstrate the capabilities of the car on challenging roads. We just didn't choose the weather, because you can't in Scotland! I picked the route that was used, and I must admit, whenever you do any car throughout your career, if you're the lead engineer on that car, you want to make the customers happy because they're the people that are going to buy that car, and also, the journalists are the ones that you're waiting on with bated breath – what is their opinion? What do they think of the car? And I remember some of the feedback at the time was, 'Amazing, how on earth have you got that ride quality and control and steering precision, and not compromised one thing or another?'"

The Cameron House Hotel on the banks of Loch Lomond suited the Lotus brand, and over 100 journalists and media guests were flown into Glasgow airport from North America, Germany, France, Spain, Portugal, Italy, Switzerland, Australia, Belgium, Denmark, Japan, and the Netherlands, and then had taken the half-hour journey to the hotel. After a quick briefing on the car, their driving experience began in the Highlands, around Loch Lomond and Glencoe – majestic or morose, depending on the fickle weather – before heading through Argyllshire down the west coast, showing off the depth of the Evora's abilities.

DYNAMIC QUEST

The March of Time ...
An interview with Lord March

Goodwood's three historic motorsport events, the Member's Meeting, the Festival of Speed, and the Revival are must-see episodes on the car-racing calendar. And that's all down to Lord Charles March one of the most important and influential figures in world motor racing; he is a huge Lotus fan – with a Gold Leaf Elise amongst his collection – and very much a true petrolhead.

I resist the temptation to floor the Evora's accelerator. I'm in Goodwood estate and I'm on the road used for the Festival of Speed hillclimb. And since I'm coming to see the patron, Lord March, I'd better behave. This is the first time I've been able to appreciate the architectural symmetry of Goodwood House, despite being a Festival regular, because today, unlike that weekend in early July when antique machinery spins its wheels, there's no giant automotive sculpture, no forest of marquees, straw bales, banners, and thousands of fans to disrupt the view of the Regency pile. Its natural setting is this English dream of expansive, verdant parkland beautiful woods, a golf course and cricket pitch. Inside, it's a grand stately home, as you'd expect, with the main reception rooms leading off the central hall – a vast area behind the portico, and beneath the balcony, as you look at the front of the house. High, decorated ceilings, tall columns, giant rugs, varnished floors, and big furniture with massive family portraits vying for wall space with Stubbs horses and Hogarth etchings.

Charles Gordon-Lennox – better known as Lord March – is running late, and as he hurries in with a friendly apology, at his behest we walk briskly down the drive to have his picture taken with his Elise and my Evora. He asks me how it drives. He's got one coming, though he's not driven one yet. His mobile rings, and he's on another mission for a few moments. This man leads a hectic life. He may have it all, but it's not without pressure. Back in the house, he steers us into the library. It's clear that every waking moment is taken up with concerns about the events Goodwood is hosting this season: The Festival of Speed, the Revival, and of course the horse racing, cricket matches and golf. There's a new attraction: The Vintage pageant of art fashion and film, mid-August; as someone who's lived with epoch-making events all his life, he's in a very good position to re-circulate

Posing with Evora and S1 Elise, Goodwood's Lord Charles March has always harboured a liking for Lotus cars. (Courtesy Antony Fraser)

them. We relax in his library with coffee and biscuits, and he recalls life as a keen 8-year-old when the house was full of heroes during one of the Easter meetings at Goodwood. Many of the drivers stayed in the house, and his grandfather Freddie March – otherwise known as Duke of Richmond – would hold a cocktail party in the house on the Saturday evening after practice, which was followed by a day off for Easter Sunday, and then the racing itself on Easter Monday. That cocktail party was young Charles's opportunity to blag signatures for his autograph book, and they weren't all happy experiences. Jackie Stewart and Jim Clark played ball but another Lotus champion was less cooperative. "Graham Hill told me to bugger off," he recalls amusedly. "Graham was holding forth in a very grand way reclining on an Elizabethan bench. I already had his autograph and I didn't really need it again but somebody pushed me in front of him so I asked anyway and he said, 'Bugger off!'"

The house is populated with memories of racing men, of early '50s ERA driver Prince Bira from Siam, and leonine Swedish privateer Jo Bonnier, sitting on the sofa right where we are now in the large library. Jackie Stewart was pointed out to him by his grandfather as being a man to watch, and the three-times F1 World Champion is a Goodwood stalwart to this day.

And then, of course, there was Colin Chapman. "Obviously, I remember him coming: he upset my grandfather quite a bit – by landing his plane here! Chapman flew in for the race, but he took no notice of air traffic instructions about what approach to take and landed where and when he felt like, and parked his plane where he liked as well!" Nevertheless, Lord March is a big Lotus fan, and followed the cars as a boy, and in fact, despite his feelings about Colin Chapman landing his plane when and wherever he felt like, his grandfather admired the cars very much. "Lotus has always been such a fantastic English brand, and has remained true to its beliefs, keeping small and clever, as with the Elise. My grandfather had a great feeling for them in that sense, because he was interested in small, clever, lightweight cars and that sort of technology."

Freddie, Duke of Richmond, was a successful racing driver himself in the 1930s, winning the Brooklands Double 12 and BARC 500-mile races in the '30s in MGs and Austin 7s, beating 4.5-litre Bentleys in the typical little-guy Chapman ethos. Bending to family pressure, he stopped racing before WW2, but he never lost enthusiasm for the sport. It was his initiative in making use of the perimeter roads of the Westhampnett RAF airfield that saw racing take off at Goodwood in 1948, when Sir Stirling Moss won the first ever race there in his 500cc F3 Cooper. During the 1950s and early '60s Goodwood was famous for several international events, including the non-championship Glover Trophy F1 race, the Nine Hours endurance race, and the RAC tourist Trophy, but a refusal to update the track to accommodate ever-quicker F1 cars led to its closure in 1966. As Lord March has it, "My grandfather went off it all in the mid-'60s – he didn't like the 3.0-litre F1 cars, he didn't like the wings and the aerodynamics that were starting to appear. He didn't like the way motor racing was going. A lot of his friends had been hurt and he didn't like being told what to do by the organising club."

At the time, it seemed to the March family (Gordon-Lennox, to

Relaxing in his library at Goodwood House, Lord March recalls historical Lotus actions involving '60s stars such as Jim Clark and Graham Hill. (Courtesy Antony Fraser)

give the proper surname) that horse racing had a brighter future than cars, as far as the Estate was concerned, and that's where they put their energy. Money was tight, and when Lord March's parents moved into the big house, it was in very bad shape and needed a lot of money spending on it. So, there was none to spare for upgrading the motor racing circuit which lay there, pretty much unused, apart from some desultory testing. But it was 'all change' with the next generation, since young Charles, the autograph-hunting motor racing fan, grew up to be a confirmed petrolhead. Though he loved bikes, his parents tried to keep him off them by giving him a Morgan 3-wheeler as his first car. "That was crazy of them!"

he says. "I bought it for £200 in Littlehampton. I was 16, and it was great – well, pretty horrible, really! It had a Ford 100E engine and it was quite quick, and quite dangerous, because the wheels regularly fell off." Still, it served the parental purpose by keeping him off two wheels, though that prohibition has long since passed into ancient history. "I've got more bikes than cars now, including a Bimota DB1 and a Ducati 888SP5, and the trials bike the children use," he happily admits. "There's even my grandfather's BMW R12. It was his original bike and there are lots of pictures of it parked outside the house in the 1930s, with his planes, bikes and cars."

Working as a still-life photographer in London in the 1980s, Lord March had more freedom to ride his bikes than he does now, when he's stuck in a suit most days. Seems a pretty high price to pay, but then the compensation is considerable. That's the joy of having your very own motor racing extravaganzas, year after year, in your very own backyard, with all those boyhood heroes and their successors, and their fabulous vehicles right there, being driven with verve and passion. So, how did that particular dream scenario come to pass? Lord March explains: "The Festival of Speed was first up, and actually it was very easy to get started. In '91 I was still working in London, and I started to look at getting the track up and running again, but it rapidly become clear that it was going to be a bit tricky. And in a slightly disconnected way, when I was President of the BARC (British Automobile Racing Club) Ian Bax, who was on the BARC board, suggested a hill-climb in front of the house."

The idea was that cars would set off one at a time speeding up the winding park road with its uphill gradient till they reached the top of the estate. With the enthusiastic support of Derek Ongaro, the track inspector for the RAC and FIA, the event got the green light, and the first Festival was scheduled for June 1993. The BARC was supportive, seeing it as a really positive initiative, though success was by no means a foregone conclusion. As Lord March says, "There were a lot of raised eyebrows that first year, because all we did was put a line of string up and people were just standing behind it. But I can remember Derek sitting in his car at the start line, and he said, 'It's no different to a rally stage, and I'm treating it no differently.'" So, but for Derek Ongaro, it all could have ended there and then. Even so, it was all still on a fairly low scale: "I tried to cover the costs with the sponsors, so that we wouldn't lose too much money." The public's appetite for the event came as quite a surprise. "We were bowled over that first year: we had 20,000 people, and we had no way of collecting the money for the tickets – we were running round with people's handbags filling them with money! I can remember lying on the grass out there, and saying, 'Oh my God, this is something that people really want!' It was a fantastic feeling." Since then it's been onwards and upwards. "When I think what

Lord March holds forth at the 2021 Goodwood Festival of Speed. (Courtesy Antony Fraser)

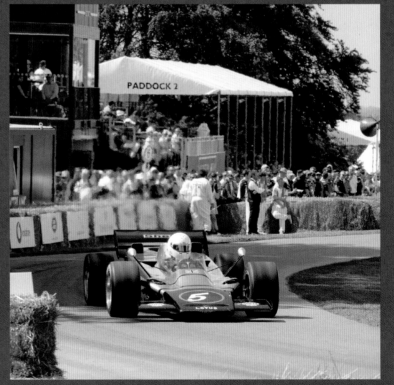

Colin Chapman's grandson, Arthur, helms the Type 56B gas turbine around Goodwood's Molecombe esses during the 2021 Festival of Speed. (Courtesy Johnny Tipler)

An Evora GT410 takes on the role of course car at the Goodwood SpeedWeek 2020 event. (Courtesy Johnny Tipler)

we spent in that first year to put the whole thing on ... we spend ten times that amount just on the loos now!" Which is why, these days, there's a contingency plan for any eventuality – even a Glastonbury-style quagmire can be absorbed and overcome.

Securing interesting cars for the run up the Hill was always a priority, and the Lotus marque was on Lord March's agenda from the outset. "When we got a Type 49 F1 car for the first time it felt like a real achievement and we certainly celebrated for a few days. That was a big moment." Eventually, in 1998, Lord March got planning permission to re-open the old Goodwood circuit and so began the Revival, the event that quickly escalated into the premier historic motor racing festival in the UK – if not the world – in terms of spectacle, pageantry, and quality of participating machinery and drivers. And, let's not forget the spectators who embrace the spirit of the occasion whatever the theme. Charles March feels that his two motoring events were established in the right order, because there was so much to be learned from the experience of the Festival of Speed. "The idea of putting the circuit back to how it was, and developing it in the way we have, came out of the Festival of Speed. From the Festival, I fortunately attracted a really good group of people who were helping and advising me, and they were adamant that the circuit needed to stay the same, because that was what was special about it. At the beginning, even my own people

Lotus has always been a prominent marque at Goodwood events, historically and currently. Here, Classic Team Lotus chief mechanic Kevin Smith readies a Type 18 for the Glover Trophy at the 2021 Revival meeting. (Courtesy Johnny Tipler)

here were a bit concerned about wearing the right clothes for the period, but interestingly, that's played a really big part in the Revival's success. After five years or so, it was evident that if you didn't make some sort of effort to join in, then you would feel quite uncomfortable and now almost everybody does."

Back to the dream, then. Because it's not just that your go-kart track has turned into a world-famous hill-climb, it's that you get to choose some of the greatest cars on earth to drive in your garden, and on your very own

As host of the Festival of Speed, Lord March gets to drive more or less whatever he likes, in this case the John Player Special 72 #5 when Lotus was featured marque in 2012. (Courtesy Antony Fraser)

race track. "The thing about the Festival is that there's always a theme, so the cars have got to fit the theme. One of this year's themes is La Carrera Panamericana, there's a selection of cars that ran in the original race series in Mexico in the early '50s." But then the fan reveals his colours: "Well, it has been quite personal, that's true. In the first few years I could just invite all the cars I'd ever dreamt of, and then try to persuade those heroes of mine to come. I never imagined I would see those cars again, let alone have them here for the weekend, and let alone be asked to drive them. The most extreme example of that for me is the Chaparral Group 7 sports prototype (winner of the 1967 BOAC 500 at Brands Hatch). I've driven the 2D and the 2H, but to drive the winged 2F, having made it as a model when I was 8 – that was a real treat …"

"Lotuses are a joy because they are small – something small with lots of power that handles like a dream, that's the essence of a great machine." Most years there's a great Lotus story at the Festival of Speed, and 2021 was no exception: the Type 56 Gas Turbine was run up the Hill by Arthur Chapman, Colin's grandson. With appropriate shenanigans, it also marked the public unveiling of the Emira and exposure of the Evija. Back in 2012, it was Jim Clark's Type 38 IndyCar which he won with in 1965. "It came over last year to be rebuilt at Classic Team Lotus, and the Ford Museum who

Lord March chauffeurs Lotus' Managing Director Matt Windle up the Hill in the Emira at the 2021 Festival of Speed. (Courtesy Antony Fraser)

own it have been brilliant. It's going to be seen running for the first time here." In fact, until CTL finished rebuilding it, mid-June, it hadn't been run at all since Jim Clark turned the ignition off in Victory Lane at Indianapolis 45 years ago, so there's a real sense of occasion about its Goodwood start-up. "It's fantastic, isn't it," he enthuses. "I'm so glad that the Festival of Speed and the Revival play a part in getting these cars known to a wider audience, and in some cases getting them restored. Owners and collectors can see that there's somewhere to take them, and that gives it a real point." I ask Lord March if he's going to be driving the Lotus 38 IndyCar, and his answer goes right back to his eight-year-old self, because he will be, but only after one Jackie Stewart has had his turn in the cockpit. "I think Jackie may have stolen my drive, actually! I am driving on the Friday but he's driving on the Saturday and Sunday."

Lord March is also crazy about big American Hot Rods, but his personal garage has hosted Austin Healeys and a Porsche 924 Carrera GT that he blew up in the Pomeroy Trophy at Silverstone. Lack of time has precluded participating in other historic races and rallies though he's done the Mille Miglia a few times, as well as the occasional demo such as Austria's Ennstal Classic rally in a Porsche 908 Spyder. It's his favourite classic event – after Goodwood naturally – and I followed his 1000-mile progress through the Italian countryside from Brescia to Rome and back in 2009 when he co-drove a prewar BMW 328 roadster loaned from their museum. He dreams of the day he can have "a nice old car" of his own so he can participate in more historic events himself. His daughter Alexandra has a 1968 Alfa Romeo Giulia Sprint GT, and plans to join her dad on the Mille Miglia sooner or later. Heartening indications that Goodwood's two classic motoring institutions are likely to have a long and happy future. As I drive the Evora away through the deserted estate, groundsmen tend the cricket pitch. It won't be like this come Festival time. The ground will be shaking, as a far more famous Lotus thunders up the Hill: courtesy of Classic Team Lotus, the Type 42 Indycar.

Lord March introduces the Emira, making its debut at the 2021 Festival of Speed. (Courtesy Antony Fraser)

Idiada – test track

Better known for its Barcelona test track facilities, Idiada became one of the most important independent specialized organisations in Europe. Providing engineering, testing and homologation services for the automotive industry. Idiada's main fields of activity are homologations, engines, emissions, noise and vibration, vehicle dynamics, fatigue and durability and passive safety. Among its passive safety engineering services, Idiada lists full-scale crash tests at EuroNCAP and FIA level, and vehicle and occupant protection simulation; dynamic impact tests and simulation for pedestrian and interior protection; sled tests including reverse acceleration and deceleration sled facilities, and simulation for restraint system development; misuse tests for airbag sensor development; analysis and optimization of bus and coach structures; design and optimization of HGV chassis and trailers.

Idiada has had a multidisciplinary accident investigation unit operating since 1999, performing statistics analysis and in-depth investigations under specific research projects with the Spanish traffic authorities, working at local, regional and national level. Matt Becker and the Evora development team made numerous visits to Idiada during the Evora genesis, and it was also on the route of my road trip from Barcelona to Montserrat in an Elise in 2009.

Passing by Idiada test facility, the route from Barcelona to Montserrat offers a succession of exhilarating mountain road bends, as experienced here in my Elise road trip in 2009. (Courtesy Jason Parnell)

CHAPTER 7
ASSEMBLY
Going down the line

The Evora was assembled on its own dedicated production line within the main Lotus factory site, originally running side-by-side with that of the Elise, so that when both cars emerged, it was from the same factory door. The width of the Evora assembly line was similar to that of the Elise line, as was the length, so the overall space employed was much the same, but because the Evora was relatively larger, the line had space to spare for future expansion. The channel along which the cars travelled whilst in-build was flanked by towering steel columns and bins crammed with the parts to be applied to the vehicle in construction at any given point.

Richard Fruin, from Manufacturing Engineering, described the modus operandi. "To start with, we had a chassis space at the beginning of the line where items like the wiring loom were installed. Then we then had the body fixing stage where the panels went on, and then we had the final trim stage. Besides that, we had two separate areas where the front and rear modules joined the main chassis tub. The chassis line was five stages long, and then the final stage is where we married the rear and front module to the main chassis, and it went onto equipment stage, and then through to the fixture stages, the first of which was bonding, where the body sides were fitted, and it passed onto a second stage, which had the doors and then the roof bonded. Then it went through to the fluid fill area for brake pipes, coolants, power steering, air-con, engine and gearbox lubricants and rolling road."

Each facet of the car passed through a specific stage, an

At each stage in the Evora assembly process an operative stamps the build sheet, confirming that the particular task has been completed. (Courtesy Lotus Cars)

assembly process proscribed in a build manual. Each element was assigned to an operative who recorded the particular action on a timer. If any process was changed, or staff were moved around, the timing could be affected, and this had to be considered along the line. There was a lot of training involved in getting the line to operate at maximum efficiency; some processes could be

Closing stages of the Evora production line, with completed but un-stickered cars to the right and the rolling road at left. (Courtesy Lotus Cars)

learned in a week, and other more complex disciplines might take three weeks of training.

Confusingly, the factory assembly areas were not numbered consecutively, so in a way the build process seemed illogical. But, of course, there was an explanation. When Colin Chapman first bought the Hethel aerodrome in 1966, the hangers that he used as factory buildings were scattered over the site. Gradually, they were united to form a coherent building plan, but they retained their old hangar names. Thus, for example, the unit that the workforce called Factory 9 was actually a building located between Factories 2 and 11. The Evora's construction proceeded like this through the buildings. The engine and gearbox assembly, and thence the rear module as well as the front sub-assembly module, were all assembled in Factory 1. These fundamental parts were then brought on trolleys to Factory 5 for uniting with the chassis and suspension assembly. The sills, roof and doors were assembled at the top end of Factory 5. The body panels were painted in the paint shop in Factory 2, and then bought to Factory 4 to be assembled, along with the chassis and suspension. It was the same for trim, which was sent over to Factory 4 from Factory 1. Factory 4 was General Assembly where all the sub-assemblies were brought together. The car then moved into Factory 8 for the rolling road, ramp inspection and final quality control. Confused? Not if you worked there ...

Clammed up

Since the late 1950s, one of Lotus' key skills has been fibreglass bodies – chassis even, in the case of the original Elite. The original Elise S1 body panels were made in-house at Hethel, but since 2000 the majority of Lotus clamshells and panels have been produced by SOTIRA. That's an acronym for Les Sociétés de Transformation Industrielle de Résines Armées, part of the SORA conglomerate, with a factory at Meslay-du-Maine in Brittany. Created in 1981, SOTIRA remains a market leader in injection and low-pressure compression moulding technologies. SOTIRA provides specialist know-how in the production of composite parts reinforced with glass or carbon-fibre. It employs a range of manufacturing processes (ICS, RTM, SMC, spray up, hand lay-up) and tooling technology at its five French plants, according to specific customer requirements regarding production rates, volumes and eventual usage. The SOTIRA conglomerate provides external and structural parts for the main European motor manufacturers, and industrial vehicles such as tractors and earth-moving equipment, plus technical and industrial solutions for a wide variety of market sectors including street furniture, railway applications, electrical and building trades.

Taking over when Lotus Cars elected to restrict in-house production of fibreglass bodywork to a few specialized products such as lightweight racing car panels, SOTIRA had the advantage of volume and capacity, not to mention relevant expertise, and used a variety of processes to make clamshells for Lotuses including the Evora, Elise, Exige, Europa and Tesla, plus ancillary panels, wings and spoilers.

Fabrication techniques

There were three different processes. What SOTIRA called 'Cannon pre-form' applies to rear clamshells and doors. The

Panels and clamshell sections for the Evora are made by SOTIRA in Mayenne, France. (Courtesy Lotus Cars)

Certain fibreglass components are still fabricated by hand in the traditional fashion, such as this Exige seat. (Courtesy Lotus Cars)

advantages were a low cycle time, typically two minutes, constant thickness of material, the facility to produce undercut lips or rims, and woven or random fibreglass mat could be used. These panels were made using the RTM (resin transfer moulding) process, basically an echo of the VARI (vacuum assisted resin injection) system that Lotus used for the Esprit. The polyester resin was injected onto a glass-fibre mat with the chrome-plated stainless-steel mould-tool hermetically sealed. A vacuum could also be created within the tool to further assist resin flow in order to avoid porosity and ensure the fibres were thoroughly impregnated.

The ICS (injection compression system) method was used to make the bits in between the clamshells, such as moulded inserts, inner wings, deck-lids and spoilers, where rigidity was crucial. Parts were pre-formed around a foam core using the RTM injection process, employing a single tool to produce both exterior and interior surfaces, and enabling the part to be painted at the same time. For ancillaries such as the windscreen hoop the low-pressure RIM polyurethane method was used. This consisted of two chemical ingredients, polyol and isocyanate, which expanded when mixed. Finally, hand lay-up implied the manual impregnation of glass-fibre matting by polyester resin in an open mould in the time-honoured method that small-volume firms like TVR, Marcos, Rochdale and Lotus were using 50 years ago. The finished product exhibited a single smooth face, and it's that which got painted. Typical items produced in this way were the grilles. Flashing was trimmed from finished panels and clamshells using CNC machines. SOTIRA had four Belotti machining centres for large parts and three trimming robots. As well as servicing Lotus bodywork requirements, and that included Elise, Exige, Europa and Tesla as well as Evora, SOTIRA also worked for Aston Martin, John Deere tractors and Unimog trucks, plus other bus, truck and plant makers.

Applying the paint
Lotus' painting process was an industry leader in respect of the media used and the booths used to employ it. The latest generation of water-based hues was in use by 2009. There was a bit of a shuffle as the trolleys festooned with body parts progressed from one end of the paint line to the other, involving at least one U-turn and a ramp or two, legacies of older spatial constraints. But there was no denying the skills of the operatives along the way, from preparation, painting, checking and rectification. My guide in the studios was another Lotus old hand, Phil Farrow, who, in consort with another supervisor, oversaw all composites, bonding, laminating, painting and checking operations. As well as the paint and body shop, Phil also ran engineering manufacturing support, plus the critical process that covered all the cleaning, and also development for colour and project work. Inception of new water-based pigments was driven by legal requirements in place in 2012, and by 2015 paint systems were completely water borne. While solvent-based paint needs solvents to flush the system, water-borne pigment requires only a small concentrate diluted with 90 per cent water. Water is the carrier

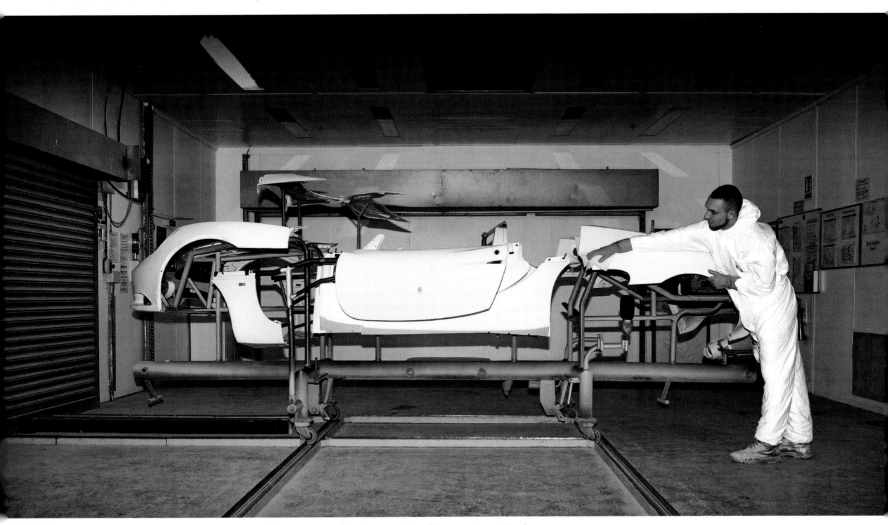

Trolleys decked with panels progress on tracks from one paint booth to another. (Courtesy Lotus Cars)

Evora panels delivered from SOTIRA are mounted as a set for each car on the trolley that takes them through the paint process. (Courtesy Lotus Cars)

Paint shop methodology involved the sprayers applying two coats to the rear clam and engine cover, two coats to the middle of the car, and similarly the front clamshell. (Courtesy Lotus Cars)

for the paint rather than solvents, which is a huge improvement on VOC's (volatile organic compounds) when compared to solvent as a carrier that evaporates. Most major OEMs (original equipment manufacturers) operate full water-based systems incorporating melamine-based paint systems with high-bake oven temperatures. "We're baking at 60-70ºC here," reported Phil. "A steel-bodied car would require a higher temperature, 120-160ºC, depending on the process." By comparison, the thickness of the paint cladding a Ford Focus probably totalled 120 microns. Lotus applied that thickness of primer alone, sanded back to 70-80 microns. With its final coat, it would be wearing 180-200 microns. Phil stressed that it is a refinish material rather than an OEM high-bake material. "The high-bake stuff goes really hard; our paint has certain flexible properties. A really hard surface is okay on a metal car, but on a fibreglass car you want a degree of pliable substance. Carbon is different again; carbon mats stay as they are, and the resin tends to grow and shrink with temperature and humidity, so the paint system has to be able to compensate for that."

Economies of scale bring advantages in different ways. "Obviously, the OEM stuff is cheaper because all the big guys, Ford, GM, Chrysler, who process 60 cars an hour, make a lot more of it, and they're using robot application processes, which again is a different technology." Lotus had the advantage of a synergy with its paint supplier. DuPont, one of the world's largest car paint manufacturers, had a technical specialist working at Lotus and gained much data through supplying and monitoring Hethel's paint systems.

Whilst the majority of Evora body panels and clamshells were painted after they arrived from SOTIRA, a minority had to be painted in the time-honoured way. "At SOTIRA, the SMC and RTM processes were very different," reflected Phil. "The way we did the Exige and Europa was ideal for a low-volume niche market production, where you'd got various awkward shapes, and you needed to produce them reasonably cheaply. We also made all the circuit racing car panels here, because their panels were much thinner: we only used two mats on the circuit car to keep it light weight." To make an Evora body side, SOTIRA used the RTM system, which consisted of two mould tools laid up with the pre-form mats. The tool was shut and injected with resin in a similar process to the VARI method pioneered by Lotus for the old Esprit. "The finish was achieved by temperature differential between the surfaces," explained Phil, "and it was this tool surface heat differential that formed a skin, and gave us something to paint onto, whereas if you painted the back of the panel it tended to soak in. It was a very effective process, and produced good quality panels." On the Elise, the majority of the panels were RTM and SMC, which was the same for Evora.

SOTIRA's panels were mounted as a set for each car on the trolley that took them through the paint process. "When we got the panels in from SOTIRA, we carried out some fettling and bonding operations," said Phil. "There would be some inspection, some filling if there were any voids or defects, certain areas where we had to fettle and cut flashing, then the guys would give it a good check over, and it would move off for a clean-up before it went into paint.

One side of the body shop was set up for Evora, the other for Elise and Exige. Parts were bonded together on jigs, set up specifically for each model, then mounted on their trolley. Phil and a colleague spent four months designing a trolley for Evora panels, rather like a heavy-duty spaceframe chassis, and they were welded up across the road from the paint shop. "It had to be fairly precise to go through the whole process, and the guys had to be happy with the positioning of all the panels so they could spray everywhere," said Phil. The component panels needed to be supported in such a way that they wouldn't distort when they went through the ovens, because they could become soft at high temperature. "The trickiest aspect was getting everything in a position where you were satisfying all the primer sprayers, all the guys that had to sand the car, and then all the colour sprayers. All us sprayers were prima-donnas, really, we had to have everything correct, otherwise we tended to throw our toys out of the pram!"

The panels moved into another bay in the rectangular tunnel, to be blown down to get rid of any dust, and washed with a mild degreaser using an abrasive Scotch pad. "It didn't cut into the panel, but it just took the shine off, gave it a slight abrasion to provide a key for the primer coat, which was the foundation for everything else." They masked around the body sides to cover up areas where they'd applied the adhesive to bond the sill to the chassis.

Phil described the painting process. "We ran with two colours of primer, because it had an effect on the top coat colour, so, for white, yellows, oranges and reds, we used a white primer, and for pretty much everything else, we used a grey primer." The primers were stirring in a pair of paint kettles, big mixing pots located in 'the kitchen.' "It was a very high-solids primer, lots of filler in it, and if you left it to settle it would just drop to the bottom like mud, so we kept it moving." When the paint was called for, it was pumped through the pipes, and when the sprayer pulled the trigger on his paint gun, it got mixed with the activator in the line, so it was activated mixed paint that came out, always fresh. This was fairly standard industry practice, though Lotus had a back-up system consisting of a pressure pot that worked on the same principle, but the paint was mixed first. This system has been operating at Lotus since the early 1990s, though the paint systems were refined to use just two different colours of primer. The system for the grey primer held about 50 litres, and they applied roughly 4 litres per car. In the primer booth, two sprayers worked one on either side of the car, working from the back towards the front, and applying two coats to the rear clam and the engine cover and then the middle of the car with two coats, and the front clamshell with two coats. The operators developed this method in conjunction with DuPont, visiting its Belgian laboratory fairly frequently, testing paints on panels in a spray booth, and seeing which guns gave best results.

The primed panels then passed into the flash-off booth for the cross-linking between activator and primer to chemically harden the paint. "In the old days, when we used to spray cellulose, it went hard because all of the solvents evaporated," Phil recalled, "whereas, more recently, the two-pack was a chemical reaction, and the solvent was only there to get the correct viscosity so you didn't need the solvent to actually make the paint cure, you needed the solvent to get the paint to atomise in order to spray it."

The paint line operated under strict humidity and temperature controls, with the primer oven's thermostat set at 70ºC and the colour oven's set at 60ºC. "They were all effectively live panels," Phil pointed out, "and they moved with the heat, especially on the Evora bumpers, which were soft flexible plastic, and they had to be fully supported in heat because they would sag, and that would affect the memory in the parts, because they wouldn't

A leopard print wrap would make an interesting livery, modelled here by Lena Lenman during a Sony camera promo. (Courtesy Phillip Pound)

spring back like the other panels. We had to have a system that would account for carbon-fibre, for GRP, RTM, SMC, ABS, mirror shells, and all the little bits and pieces that were made from different materials that reacted differently to paint and to temperature."

All the bodyshop and paint shop operators had an acute awareness of what they were making. They needed to be sensitive, because the Evora was a different car to rub down, compared with an Elise or Exige. Carbon-fibre had its own particular issues, as it could be prone to surface issues, perhaps where some of the resin hadn't been fused into the carbon, and any carbon strands had to be repaired, which was expensive and time-consuming. It would be the same if there was a split in a rim bumper. "The sanding deck was one of the first places we started people off on, because it was a basic process," said Phil. "It was about understanding the limitations of the product, basically, so the guys would start here and then move onto inspection, and as they got more proficient and experienced, then eventually they'd build up, and some of them might become sprayers, some of them worked for me, some became Team Leaders. One of the guys who was one of my supervision colleagues started on the deck, and in seven years he had gone through the whole shop, so the knowledge that he built up allowed him to run this area."

Air-conditioned chill-out areas enabled operatives to take a break, though all employees were forbidden to touch anything with bare hands because it was the start of the painting process, and hence all the primer painters and sanders wore gloves when they touched the cars. The basic gun-finish primer was quite a smooth surface, with a reasonably high build. Then came the first checkpoint. With the white-primer cars the operators dusted the panels with a black powder, which revealed any defects, holes, paint runs, which would call for minor rectification with filler and a spot repair for primer. By the time it was smooth, they knew they'd got rid of all the structure of the primer, along with any imperfections. They didn't use the powder on the grey primer, because structure and imperfections were more easily visible. Edges were most susceptible due to the use of machine sanders, so they tried to avoid using them on panel edges and feature lines. Instead, these were sanded using soft hand pads.

Next up came the Ballroom, so-called because the panel sets were shuffled around in order to get the optimum line-up in terms of colour specifications on any particular day.

Before the application of the finish coat there were interim procedures to fulfil. The cars passed along the tunnel into the blow-down chamber, where the panel set received a comprehensive dusting by air, and then proceeded to a wash station, where a water-borne degreaser was used to clean the surface of the panel set, in contrast to the primer coat's solvent degreaser, to remove any contaminants.

Enter the shooters. "Those paint guns could be tuned to almost a pencil point," declared Phil, "so you were spraying at low pressure, and getting a very fine spray, almost air-brushing in any exposed substrate on the edges. We'd put around 5 microns of a spot repair air-dry primer coat on, just enough to cover. This stuff dried within five minutes. It gave us a homogenous colour over the whole car, and was quite ideal because it dried so quickly, and it didn't have to be sanded, you could paint straight over it, because it was basically a base coat material formulated to match the primer." At this point, the body set performed an intricate U-turn, passing through pneumatically sealed doors and onto a ramp that elevated it into the final painting arena. When an alarm sounded to indicate that the particular stage in the process had been completed, the car would then move into the colour booth. Here, a tad alarmingly, the floor consisted of a steel grille with water swirling inches beneath the surface. There was an imperceptible downdraft in the booth so the overspray drifted down into it. "The water was only about 6 inches deep," said Phil, reassuringly, "and there were large trays underneath, and a chamber called a hydro-pack system. Chemicals in the water caught the paint particles and took them to the bottom section, where they formed a sort of raft and got skimmed off, and the water re-circulated. When the booths were cleaned out during the summer shut-down, the paint residue in the bottom of the trough was as thick as mud, and had to be shovelled out."

The base-coat spray guns were the latest Iwata models, with controls to adjust the fan pattern and the volume of base coat. A needle ran through the fluid tip, and the trigger pulled it back and released air through the gun, which atomised the paint jet. Obviously, there was a knack to it. "Spraying was a lot about

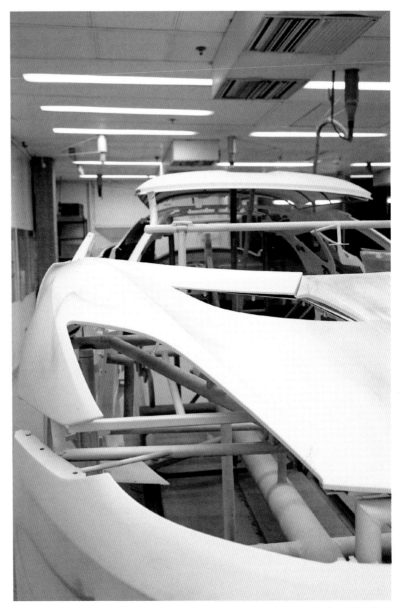

Painting was a complicated process: after priming, panel sets were dusted with powder to reveal defects and, after rectification, received a blow down, followed by a water-borne degreasing prior to top coat.(Courtesy Lotus Cars)

the feel," said Phil. 'When you sprayed a car, you were looking at what happened to the paint as it hit the panel, and you just got a feeling of what you needed to do. Lots of little globules of paint were hitting the panel where they merged into one another, and there were a lot of factors in play: there was air pressure, paint viscosity, there was temperature, humidity, airflow: all these things had an impact, but basically you got a feel for it. You moved with the car, it was almost like you caressed the surface; you were dancing with it, in a way."

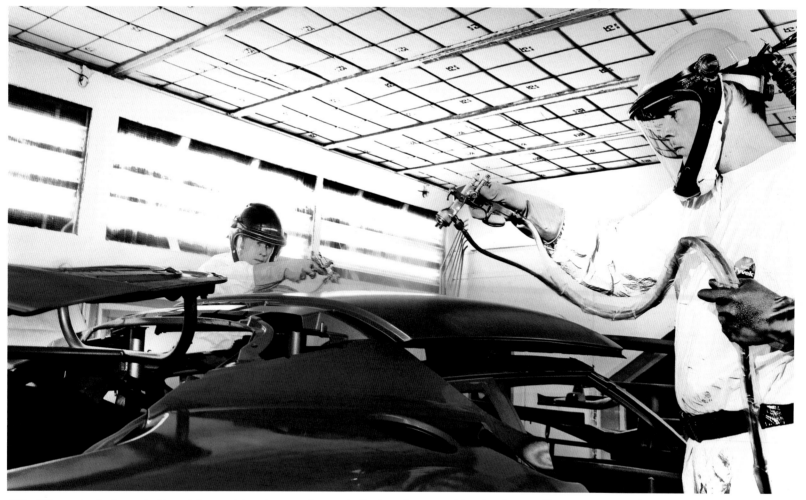
Attired like a pair of astronauts, the painters work in unison, moving gradually along the car applying the pigment. (Courtesy Lotus Cars)

Two painters arrived, garbed like astronauts, and started to spray, working from the back of the car. Their action was methodical, calculated, swaying this way and that, training their guns on the panels, and they gradually turned from white to yellow. The sprayers did the edges first because on some of the sharper contours the paint could go a little bit thin, then they started to cover the upper casings. When running on volume at 80 to 100 cars a week, there would probably have been four base coat painters at work. The two painters stood on trestles beside the rig and applied two coats, spraying one area at a time, like one wing. As with the primer coat, they started at the back and worked their way to the front. For the most part, the inside of the shells stayed unpainted as the majority was unseen, or covered with a wheelarch liner. Painting different areas of the opposite sides of the car, one aimed high while the other ducked low. "We got it down to about 16 minutes with four guys when we were running 120 cars a week," remarked Phil. The painted panels moved along into the flash-off area, where temperature and humidity monitors logged conditions four times a day. "It needed to be quite warm in there, about 30-35 degrees and about 20-25 per cent humidity," said Phil. "Continuous airflow dried the water off the paint, and you got a nice, flat, even finish on the paint." Once it had gone dull, it was dry and ready for the clear coat, so then the rig moved through to the next bay for the clear coat, which was the shiny lacquer. The build sheet for each car went along with it, protected in a tube. "It told us the spec of the car, and our paint mix guy referred to it, so he'd always got the right paint mixed up a day in advance." This was specific to the colour people, so once it had been sprayed they would tick the box. The clear lacquer coat was solvent-based. "It was like the primer; you mixed it with an activator that went off, and it became a hard, shiny shell."

The colour range was 22 colours in the early days, but now can be as vast as you can imagine. DuPont has some 85-90,000

Operators administer a buffing to painted panels. (Courtesy Lotus Cars)

Final polishing as the car approaches Lotus Buy-Off. (Courtesy Lotus Cars)

formulae to choose from, and new hues are always under development. "Silver was always a popular colour," said Phil, "and stormy titanium colours, various shades of metallic greys. And white made a come-back, with a pearly effect as well. Yellow was always a good colour, and the Evora looked great in Canyon Red. Black, metallic black especially, greens and blues, blues were usually good; we tended to have a light blue, mid-blue and a dark blue. I worked closely with the Design Studio and sales teams on colours. We tried to mix the colour range up each year with regard to the general colours, but variations of them and then maybe a couple of new more extreme colours. We brought out three new colours that suited the Evora, which were more reserved, subtler, but classier: carbon grey, aquamarine and quartz silver."

Morning assembly

Once panel manufacture and painting procedures were in hand, the modular build process could begin. Starting at the top end of Factory 5, Stage 1 involved installing the main wiring harness in the bare chassis, and the VIN stamp riveted on. This denoted each car's specific ID, engine size, model year and where made. At Stage 2, the fuel tank was put in, followed by the steering column, and then the pedal box, gearshift and

As the central chassis sections start their journey along the assembly line, the first stage involves installing the main wiring harness, and the VIN plate is riveted on. (Courtesy Lotus Cars)

The nucleus of the cabin, the wiring loom and part of the pipework have been installed in the first stage of chassis assembly. (Courtesy Lotus Cars)

Front and rear chassis sub-sections were built up in Factory 1 before being united on the production line with the central chassis. (Courtesy Lotus Cars)

handbrake assembly. Next, the steel pipes that carried fluids along inside the chassis were inserted, having been made up on jigs. Station Red 400 was engaged in fitting the backboard and the rear bulkhead glass, plus the main seatbelt frame and the main roll bar. The ABS modulator, brake pipes and wiring up of the harness along the front bulkhead were next in line.

The line then progressed to Red 500, which was where the rear subframe was attached. This came in, ready-fabricated, from Factory 1, and the operators hoisted it onto a frame and brought it over to the line. It was bolted in as one big module, complete with driveshafts, wishbones and brakes. The second stage of Red 500 was, logically, the front module attachment, mounted by the same operators, and incorporating steering rack, radiator and H-vac. The heating controls, brake discs and front crash structure were bolted onto the front module. At Red 600, the front-end plumbing included the cooling pipes and AC wiring. This was also a QIP (quality inspection process) stage, and an operator examined the chassis against a check sheet. Any problems discovered were rectified straight away.

The rear module of the Evora was sub-assembled in Factory 1. After some machining work on the base engine unit and gearbox, freshly arrived from Toyota, the powertrain's two main components were mated up. The steel rear subframe was manufactured in Lotus' own foundry off Vulcan Road on the industrial estate close to Norwich airport; then it was sent off to Colchester to be zinc-galvanised before arriving at Hethel. Here, the team prepared the subframe before loading it with the engine, including positioning the handbrake cables and some suspension components: tasks easier to achieve with the engine out. Once the ancillaries were on, the team dropped the engine in

Operators fettle front and rear sub-frames, while hub carriers, suspension uprights and brake components are assembled. (Courtesy Lotus Cars)

place with a special crane and fastened it in position. Then the operators finished off the suspension, fitted the uprights and the brake discs, assembled the calipers and the exhaust system, and some of the plumbing on the engine bay. The complete module was then transferred over to Factory 5, where it was bolted straight onto the back of the chassis, and cables and wires connected up.

Seven stages of assembly were accomplished in this area of Factory 1: three on the engine, and four on the rear subframe. The assembly area for the front-end of the car was nearby, comprising the aluminium subframe received from Lotus Lightweight Structures. It was put together in a similar way to

The wishbones and coil-over dampers are fitted during build-up of the rear chassis module. (Courtesy Lotus Cars)

The base engine unit and gearbox are wheeled into Factory 1 for sub-assembly, where the powertrain's two main constituents are mated. (Courtesy Lotus Cars)

the rear end, including the cooling radiator, the air-conditioning and heater unit, and the steering rack. Then the operators assembled the wishbone suspension, so that the front module was a complete assembly ready to be added to the front of the main chassis. Each front and rear subframe assembly took around eight hours to assemble, before the modules headed over to Factory 5. The engine was canted over at an angle, so ancillaries were still accessible. All servicing points were readily to hand, which was helpful to the operators. Another piece of equipment to aid installation dropped the radiator into position, then held it while the team added the bracketry, also holding the heater assembly in place whilst the bolts were inserted.

The operators in Factory 1 added their paperwork to the build manual and the build book. Quality inspection was carried out here as well, so that when the sub-assembly arrived on station, the build team in Factory 5 could be sure that everything was up to standard.

The body panels could now be attached to the completed chassis. Each jig represented a stage. The first four stages were carried out on rollover trolleys, so if the operator needed to work in a horizontal or vertical plane, the trolley could be flipped accordingly. Then the sub-assembly was moved up to the bonding trolley at Red 700, the bonding stage, with sub-assembly benches either side to accommodate the body sides,

ASSEMBLY

Synchronised jigging: matching jigs ease the body sides and roof up to the central chassis, where they are bonded onto the cabin module. (Courtesy Lotus Cars)

The roof and door frames are placed by jigs onto the central body section. (Courtesy Lotus Cars)

Operators bond the roof in place on the windscreen header and door frames. (Courtesy Lotus Cars)

The central cabin section of the bodywork consists of roof, windscreen surround, rear three-quarters and doorframes; doors are fitted next. (Courtesy Lotus Cars)

doors and roof during their attachment. The protective tape was then removed, the panel cleaned, surface primed and glued with twin-pack adhesive, having been dry-fitted first to make sure everything lined up accurately. The bonds were cured under heat before the chassis moved on to the next process. The panels were mounted on a pre-set jig, which slid in to the chassis, ensuring the body sides were in correct alignment. Twenty minutes later, when the heaters had cured the adhesive, the assembly moved onto the next stage, which was to fit the roof. This was bonded in place in the same way as the door frames. Suction cups were applied to the panel surface, the roof was hoisted up, carried over to the correct location above the car, clamped on and held for another 20 minutes while the curing took place.

The doors were built up in a sub-assembly area and transported on trolleys so they could be installed without any manual lifting. The door sub-assemblies employed specially-designed jigs for ease of fitting them to the cabin, which could be tilted as the operator required, controlling their orientation by means of a foot pedal. The glass, door regulators, window mechanisms and electrics were all incorporated at this stage. Each inner door panel arrived in six separate pieces, which were assembled on a bench, and then the operators offered the panel up to the door and fixed it in place.

The Red 900 work station contained a two-post ramp, because it involved operators working underneath the car. So, the chassis was lifted into the air for handbrake fit and sheer panel fit – that's the structural panel covering the fuel tank and underbody area. QIP operators would check the advancing chassis at this stage too. At the Red 900 stage, the rear quarter-light glass, the front windscreen, washer bottle and heating ducts for the front of the car and the front module were incorporated. After this stage, the nascent Evora was dispatched to Factory 4 for fluid fill. In just 100 metres, a bare chassis became a rolling chassis, all bar the wheels.

The production line originally ran in parallel with the Elise, so Stage 100 was mirrored on both lines, as was QIP. Next door in Factory 4 was General Assembly. Here, fluids including brake and clutch, oil, water air-con and other lubricants were run through the complete system. The brakes, fuel system and AC were all pre-tested before the fluids were put in, so in theory there would be no fluid leakages. If a component failed a prior pressure test, the machine would not inject any fluid in, at which

At Red 900 work station the build is well advanced, with handbrake, undertray, rear quarter-light glass, windscreen, washer bottle and heating ducts fitted. (Courtesy Lotus Cars)

ASSEMBLY

Two operators using suckers carry the Evora windscreen into place. (Courtesy Lotus Cars)

Evora steering wheels with their distinctive flattened lower ring are made and trimmed in-house: here awaiting fitment along with IPS shift modules. (Courtesy Lotus Cars)

The bare chassis becomes a rolling chassis, all bar the wheels, in just 100 metres. (Courtesy Lotus Cars)

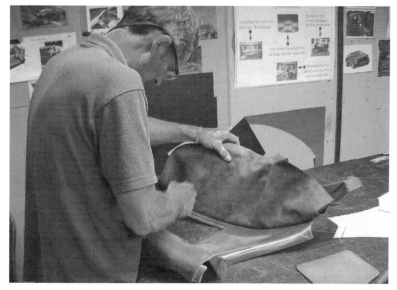
Trimming hides for cabin upholstery during the development phase. (Courtesy Lotus Cars)

point the operative had to find the leak. This only happened very rarely, and the pressure was set at a higher level than the system required. The process of sub-assembly involved installing the steering wheel, lights and side markers. With the trim panels sub-assembled, it required minimal work inside the car.

The trim shop

Factory 1 was where Colin Chapman started in 1966, and was where Lotus built the Esprit, Excel, and M100 Elan. As new buildings went up, it became a storage area, but in 2002, the trim shop was relocated from Ketteringham Hall and installed in Factory 1. This spacious, hangar-sized environment was a busy place, revamped to include an engineering workshop to run the Evora press fleet and sustain the car's development, as well as the testing programme located on the other side of a dividing wall from the trim shop.

The Evora's trimming process was just as interesting and involved as mechanical and bodywork. It was broken down

Sewing the Alcantara cladding on the Evora steering-wheel rim. (Courtesy Lotus Cars)

ASSEMBLY

Marking up the hide to indicate potential blemishes in the leather. (Courtesy Lotus Cars)

into small elements in much the same way as the line stages for the build of the car. Manufacturing engineer Pete Hunter joined Lotus in 1994 and worked in the paint shop till becoming special projects supervisor on the Evora team in 2007, with a brief to source the upholstery and trim. Having visited Bentley and Connolly, who supplied trim for Aston Martin, he realised that Lotus had a smaller skills base, and that the work could be brought on site. The traditional system of hand-cutting leather panels was very time-consuming, so Lotus bought in a massive computerised leather-cutting machine. As Pete commented, "All you had to do was mark up the imperfections in the hide which you wanted it to avoid. We'd already digitised our patterns into the computer, and it would then orientate the hide in the way which would get the best fit."

Unrolling tartan cloth for trimming Evora cabin upholstery. (Courtesy Lotus Cars)

At that point in time, the machine was run by Sandra Secker, who marked out the hide with different colours: blue crayon for an area that couldn't be used at all, and green for the remaining parts. A vacuum sucked the leather down onto the vast table, and then Sandra applied cellophane to keep the hide tight so that the cut was not distorted. A camera scanned the hide, picked up the flaws, and the patterns were then applied to the spread-out leather. CAD software digitised the patterns, which were all logged in a database also housing those for other Lotus models, so that that they could be cut out. The CAD screen illustrated how the hide would be applied in the car's cabin interior, so that the flaws were hidden and the best fit applied for the cut out. If production volume climbed to its maximum, the machine could be operated by two people to cut one hide every five minutes. Pete reckoned that, without the new machine, he would need to employ an extra five cutters, taking up much more space. Twenty-three staff operated the trim shop, occupying three-quarters of the building. Investment in new equipment extended to new sewing machines as well.

Although Evora seat frames were made and upholstered by Recaro, the actual hides used were provided by Lotus. (Courtesy Lotus Cars)

Cutting the carbon-fibre sheet for the manufacture of trim panels. (Courtesy Lotus Cars)

The stylists in the Design Studio selected the colours, and the hides were supplied by Andrew Muirhead & Son, an old, respected Glasgow firm. "Our materials comprised five different types of leather, ranging from Sateen to Arisaig, which was three times as expensive. We had to conform to automotive industry fire standards, so they had to be fire retardant. We used the whole hide, apart from the belly, which was too grainy, and the back, as that's where the bones would stick out. So, it was mainly from around the beast's shoulders and the main body." The Evora colour range was five different colours, either all black, or the base colour in black and seats in paprika, chestnut, grey or oyster, and the band that wrapped around the cabin matched the seats. "We didn't use Alcantara any longer, because that was very expensive; instead, we used a cloth called microfibre, which looked to the outside world like Alcantara."

There was a schedule carefully defined by the build sheet. "When a particular car went through to the paint shop we would get the hides out, mark up the defects on them, cut the patterns, sew them, and then when the car reached Factory 5 they installed the seats. It gave us about a day." While Elise and Exige seats were upholstered in-house, Evora seat frames were made by Recaro in Germany, and the seats came from Recaro ready-trimmed. "We cut the leather for the Evora seats on our machine at Hethel, because the leather that Recaro used was Volkswagen spec, which was very expensive and didn't match our leathers. So, we cut out our Sateen leather and sent it there, and we would make sure they had a stock of it for the Evora covers. Recaro held

ASSEMBLY

Trimmers cover door and trim panels with Alcantara. (Courtesy Lotus Cars)

A machinist sews upholstery segments in the trim shop. (Courtesy Lotus Cars)

Seats like these monogrammed tartan examples in a GT410 are bought in from Recaro in Germany. (Courtesy Lotus Cars)

about six week's-worth of stock, and we ordered about two weeks in advance of the car going onto the assembly line."

The carpet used in the Evora was a fabric common to the automotive industry, and cut to pattern. The luggage boot was trimmed in-house, as were the tailgate liner and the stowage net. Lotus trimmers clad the dash, which accounted for seven pieces of leather with another eight pieces underneath. The main instrument panel fascias were all leather, while the door panels consisted of five separate parts accounting for another ten pieces of leather. Gear knobs and handbrake gaiters were in leather, totalling some 36 pieces. As Pete said, "Depending on what variations you had, whether it was a left- or right-hand drive cabin, that is a lot of different parts. In a two-plus-two, there were different rear quarters to allow for the seatbelts to pass through, so the configuration was quite different to a two-seater. Time-wise, it was fairly labour-intensive. For instance, it took about 90 minutes to make a set of door panels, and about six hours to make a dash, as there were so many parts and an awful lot of work on it. In total it was probably 30-31 hours to completely trim a car."

The contributory leather parts were single-stitched and pinned, and then handed over to be applied to one of the different and often intricate trim mouldings that together made up the Evora interior. One of the upholsterers sewing the leather patterns, Tracey Cushing, told me, "Once upon a time you could have any colour you wanted, as long as it was black, and red and white were as far as variations went, but with the Evora it could be whatever the customers asked for." She cited pumpkin and blue, and a local radio station's promo car in fuchsia pink and lime green. Customers were given a book of samples for inspiration, and the personal service extended to embroidering the leather with initials, or weaving names into the carpet. The array of different specifications was bewildering, as the cabin and luggage areas required nearly 100 different pieces to complete the upholstery. It has to be said, though, that with the advent of the Emira in 2021, Lotus has outsourced much of the trim shop, so the trimmers were being relocated to other areas of assembly.

Rolling road

The final build stage involved the car going back on the ramp, where the underbody check was repeated, the air-con and electrics were tested and, finally, it was then fuelled up to be classed as finished. However, a host of checks were yet to be undergone. The rolling road was a booth within Factory 8, and every car was put onto it to double-check brakes, speed sensors, and air pressure (TPMS) sensors on the wheels. The alarm systems were checked and set. Each aspect had to pass these checks and receive a certificate, and any failures were returned to the factory to be fixed. As a car entered the rolling road booth, it was lifted onto the rollers to run both sets of wheels. They were double-checked, and the VIN number applied. All the data was locked into the EMS system, so that the running information on each particular car was locked into the engine's brain. This set the fuel, emissions levels and sensors. The rolling road performed all the electrical checks, covering ABS, air-conditioning, fan speeds, electric windows and lights. It verified the engine mapping and ECU programmes, ensuring that the car was, electrically, fully functional.

The rolling road, an Actemium unit common to most motor manufacturers, was a significant productivity gain, since just a few years ago Lotus tested every car on the track, barely 100 yards away, all very well during the development and evaluation periods, but hardly state of the art for sign-off. The operator could run the car as if it was out on the track, assess speeds and horsepower outputs, monitor emissions, check cruise control, apply sport mode: the rolling road placed all those criteria in one location, and indoors within the factory. Having finished on the rolling road the cars were put on a ramp for an inspection of the undertray, because they would have been run hot, all the fuel lines and oil pipes were double-checked for leaks. Each car spent 20 minutes on the rolling road, and 20 minutes on the ramp, making 12 cars going through in a normal eight-hour working day.

The Evora's cycle was a comparatively long one. The base-model Elise, with no air-con or TPMS, required fewer checks,

Having reached the end of the assembly line, an Evora GT410 receives a last buffing during final inspection. (Courtesy Lotus Cars)

Applying the Lotus lettering stickers to a completed car. (Courtesy Lotus Cars)

so it took between 12 and 15 minutes to pass through, and the Europa took 15 to 18, while the Evora needed 20 to 25, being a fully-laden car. The Evoras then went onto the Hunter ramp to have their suspension geometry checked. This would normally be done in Factory 5, in line-build, but the Evora, as a higher-grade car, was done off-line, which allowed for more flexibility in setting up a car to a specific series of customer requirements in terms of suspension settings. Next step for the Evora was down to what they called paint finals, where all the paintwork was double-checked for the minutest blemishes. When the cars were complete and had passed all the checks, every single one was water-tested in a booth close to the rolling road.

At this point, the cars were driven in for their engineering and GCA evaluation. Colin Matthews of the Quality Department explained what took place here: "GCA is the global customer audit, and all the information it produces is fed back into the build process." The GCA, which took about four hours, was based on a General Motors system of scoring the quality of the paintwork, the build-quality and the car's driveability. Two or three cars were singled out at random each week, parked in the yard and then subjected to this detailed inspection. Each factor or defect on the car was scored at 0.5, 1, 10, 20 and 50. So, for instance, each water leak counted as 10 penalty points, wherever it was found that could be obtrusive to a customer, in the boot or in the cabin. One paint dribble counted as a 10, while missing off a legal label scored a hefty 50. The base line score for the whole car was 20, so anything above 20 generated work for supervisors and the workforce to cure the problems. Anything below 20, and the car was deemed to be acceptable for delivery to a customer.

It was difficult to achieve a routine 20, and Lotus found on benchmarking competitor cars that only one rival manufacturer, Porsche, was reliably under the 20 mark. GCA over, the car goes down into the paint shop again for polishing and finishing before it comes back to be sent off down the LBO line – that's 'Lotus Buy-Off' – a final defect audit to check that nothing was loose, everything fundamentally worked, and the build books were all correct and up to date. The build books are an essential auditing resource for each car's build history. Every vehicle at every build stage in every area has to be stamped by the operator to say it has all been done correctly. The numbers in the book are job numbers, so Lotus would know exactly who'd built the car and when it was built. The book logged all the QIP stations the car had been through, so that everything identified as needing attention could be seen to have been fixed and checked off. Any element damaged on line, or anything needing to be rectified by the paint shop was all noted in the build book.

During the Evora manufacturing period, each factory stage employed an average of four people, each of whom was obliged to sign off the build book. These books were archived in Factory 3, so that five or ten years down the line, a customer buying a

Towards the end of the line, complete cars were checked for water-tightness in a booth close to the rolling road. (Courtesy Lotus Cars)

secondhand Evora could ascertain precisely what happened to the car during the build process. Every single component supplier is identified, so any changes in supplier can be tracked, as each change is noted as a VIN change. If a customer orders a car with a bespoke specification, a deviation request form is filled out and signed off by Manufacturing Engineering, a line supervisor, Type Approval and Material Control, to make sure that whatever is specified for that car is legal, so when the car is built to that specification the build book will incorporate that as well.

Lotus Buy-Off was where the rest of the foot mats were installed, the interior was cleaned up, and the car was then, in theory, ready to go off to a dealer. The pre-delivery inspection cost the dealer £60 per car, since the dealer had almost nothing to do to it, as the car was pretty much exactly as the customer would receive it. It was also where the Lotus-officiated form for the alarm was applied, so that the buyer could shop around for cheaper insurance by having the correct immobilisers.

Lotus transformed its build quality over the 10 years leading up to the Evora launch, a process that became increasingly refined throughout the model's lifetime, achieved through a combination of ergonomics, quality management and workforce training. The proof of the pudding, as they say, or in this case, is in the driving experience.

LOTUS EVORA

Completed Lotus models are subjected to the GCA (global customer audit), which ensures there are no blemishes. (Courtesy Lotus Cars)

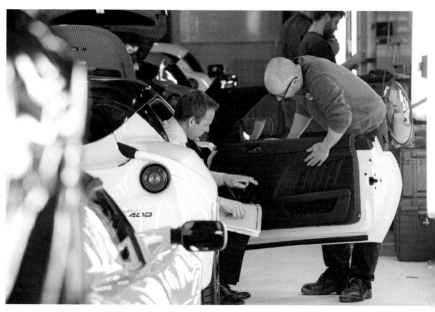

Checking a finished Evora 400 to confirm it complied with the GCA evaluation took over three hours. (Courtesy Lotus Cars)

End of the line: completed cars in Lotus Buy-Off receive a final cleaning and have floor mats installed: they're then ready for delivery. (Courtesy Lotus Cars)

CHAPTER 8
ROAD TRIPPING
Driving thrills

Hustling the Evora around the Picos de Europa hairpins, passing through Asturias en route to Portugal. (Courtesy Antony Fraser)

Drive stories are an essential aspect of any car book or magazine, and my personal forté is to take a vehicle from factory floor, dealer showroom or garage forecourt, and turn it into an animate object, bringing it to life by describing what it's actually like to sit in, to control from behind the wheel, describing how it handles and delivers its performance. Obviously, these interpretations need a little geographical context, and here's where I transport you to some fairly exotic places, in an exciting and edifying way, learning how the car is behaving en route.

To Verona and the opera – by Evora

As the only mid-engined two-plus-two Grand Touring car in the world, the Evora fits comfortably in an Italian tradition of similarly configured Ferrari and Maserati models; paying homage to those masterpieces, I drove an Evora to Verona, calling in at a few iconic Italian locations such as Monza along the way.

The 1000-mile (1609km) trip from the Hethel factory to northern Italy's spectacular Lakes took two days from the UK, travelling through the Swiss Alps via the St Gotthard pass. In consort with Lotus Club International's Caroline Parker and in-house photographer Jason Parnell in their Peugeot rent-a-car, me in the Evora, I approached Como in fairly heavy traffic. As in, a snail's pace. Entering a roundabout at a crawl, I was alarmed to see a small Fiat approaching from my left-hand side alarmingly quickly. It sideswiped the Evora with a graunching crump, and shards of broken Lotus flew in the air. The Fiat came to a standstill a few metres away, and an elderly woman got out. Shell-shocked, I sat still in the Evora, unable to believe my ill fortune; the car was probably wrecked and we hadn't even taken a single photo of it yet. The woman fell to her knees, hands clasped heavenward in supplication, presumably imagining she had killed me.

In no time, the Lotus was surrounded by jabbering Italian motorists who took the old woman's side and berated me – the bloke in the sports car – for causing the accident. A minute or two later, Jason and Caroline showed up, providing moral support, and it wasn't long before the cops arrived too. Nothing to see here, they concluded, and took off, and when it was clear that no arrests would be made, the rest of the spectators drifted away too. The old woman's Fiat was trailered from the scene, but the Evora was apparently in a bad state, with cracked nose panel, cockeyed front lid and misaligned nearside headlamp, and split front nearside wing. Lotus Bergamo to the rescue! Patron Stefano d'Aste, who runs PB Racing and is a prime mover in the Italian and international GT3 and GT4 Championships with his Exige and 2-Eleven, sent a truck to collect the hapless Evora. By close-of-play the following day, his mechanics and technicians had realigned the suspension and tracking, as well as straightening and patching up the broken bodywork. From Jason's subsequent photos, you would never know it had been involved in any kind of altercation.

We were based just out of Verona at the erotically-charged Biblos Art Hotel Villa Amista, with its larger-than-life nude wall

Stairway to heaven? The Evora parked beside the erotically-charged Byblos Art Hotel Villa Amistà, between Verona and Lake Garda. (Courtesy Jason Parnell)

paintings and flamboyant décor and furnishings. Closest of the Italian Lakes was Garda, 15km away. Clear, placid water, with an Alpine backdrop, ranges overlaying one another and receding into misty distance. Down by the water's edge, a sinuous two-lane highway tracks every nuance of the lakeside, flanked by spindly cypresses, umbrella pines underplanted with camellias and azaleas. The road bisects towns and villages: Bardolino, Lazise, Tori del Benaco and Torbole, and Garda itself, where villas are painted chalk white, yellow ochre and rich red-brown, the more flamboyant decorated with frescoes. Frequent lay-bys and quaysides give access to a slim strip of pebble beach, where rudimentary stone promontories and jetties harbour tiny fishing boats bobbing on sparkling blue water. Fishing was once the driving force of the local economy since there are trout, turbot and eel aplenty in Lake Garda, but now tourism rules. Paddle

On the quay at Garda: the Alps and the Dolomites provide a backdrop to the Lake's northern shores. (Courtesy Jason Parnell)

steamers and catamarans ferry trippers from one side to the other – it can take a day to circumnavigate the lake by road due to traffic and local distractions. Chocolate-box Desenzano is a yachting centre, while little medieval Sirmione with its moated castle is on the end of a spit extending far into the lake at its southern end, giving the impression of being out on an island.

By mid-morning, holiday traffic drives us away from the Lake shore and up into the range of mountains to the east – Monte Baldo and the Parco Regionale – where there are wonderful hairpin-punctuated lanes and short, fast straights, winding ever upward through the woods, barely wide enough for two cars. The Evora's rasping V6 echoes determinedly off the cliffs and stonewalls, causing the occupants of a posse of Beemers coming downhill to stare open mouthed as we rumble past. The asphalt is generally in good heart, though the Evora rides bumps with ease in any case. As we head east, the hillsides are terraced with vineyards: that's because we are in Valpolicella territory. Light and fragrant, this fruity tipple comes exclusively from the region north of the Adige River between Verona and Lake Garda, and it's almost as prolific as Tuscan Chianti. The grapes it's made from are Corvina Veronese (40-70 per cent), Rondinella (20-40 per cent) and Molinara (5-20 per cent), mainly seen growing in pendulous clusters on vines sprawling on overhead pergolas. A higher-quality blend is identified as Valpolicella Classico Superiore, while Reicioto della Valpolicella is one of Italy's top wines. Bardolino is another easy-going table red from the shores of Lake Garda, which has its own wine festival, the Festa dell'Uva, in late September, while Soave is the region's dry white. Bardolino also has an olive oil museum.

Marble quarries abound, fundamental to the classical splendours designed by leading Renaissance architect Andrea Palladio. While cherry orchards deck the hillsides, the descent to

Rushing the Evora towards Monte Baldo and the Parco Regionale, winding upwards along short, fast straights and wonderful hairpin-punctuated backroads. (Courtesy Jason Parnell)

the corn-basket plains of the Veneto is marked by a transition to arable farming, maize fields alternating with wheat, though strong aromas of cattle waft through the humid air. Even in late June the combines have started the wheat harvest. Avenues of Lombardy poplars, little canals and aqueducts define the landscape, with more orchards, vineyards and big wine co-ops. The appropriate weapon of choice on these field-boundary farm roads would be an Elise, though the bigger Evora acquits itself well enough. Birdsong is audible in quiet moments, but the chirruping of cicadas is omnipresent.

That evening we keep a rendezvous in Verona, not with Romeo and Juliet, nor the Two Gentlemen of that ilk, but with Aida, Giuseppe Verdi's protégé who's playing under the stars

The smooth blacktop flanking Lake Garda was irresistible for tracking shots during our Hethel to Verona road trip. (Courtesy Jason Parnell)

The SR249 runs alongside Lake Garda for 50km, from Peschiera del Garda in the south to Riva del Garda at the top end: the Evora waived the laps. (Courtesy Jason Parnell)

at the vast outdoor Arena theatre. The Romans who built this coliseum in 30AD knew a thing or two about acoustics, using amphorae to amplify the sound, and while those terracotta pots might not now be in evidence, the fact that the opera is broadcast with no microphones or sound system is testament to the Roman architectural ingenuity. It's the third largest of all the Roman amphitheatres, originally capable of accommodating 30,000 people. At 46m, almost the width of a football pitch, it's as broad a stage as any thespian or operista could possibly wish for, interspersed for the set of Aida with vast columns, numerous sphinxes, temple and palace architecture, accommodating the cast of Ethiopian and Egyptian royalty, 100 soldiers and half as many priests, plus 30 or so diaphanous nightie-clad dancing girls with no problem. In the melodramatic love-triangle, Daniella Dessi is slave-girl Aida, Fabio Armillato the Egyptian commander Radames, and Tichina Vaughn the jealous princess Amneris,

daughter of the Pharaoh. It is absolutely spellbinding and we are mesmerised by the visual and aural spectacle. Aida was the last opera Verdi composed, in 1871, and originally staged at the Arena in 1913. It's the most oft performed of all operas with 542 performances since then – Carmen is next up with 193 shows.

Like the majority of provincial cities in north Italy – Brescia, Vicenza and Padua – the medieval heart of Verona is crisscrossed with a maze of lanes, just wide enough for the Evora, but it's cycling pace much of the time as I nose the Lotus through the crowded confines to the Piazza del Erbe, the city's produce market with its parasolled street cafés and renaissance fresco'd buildings. Great artists who worked in this theatre of excellence over the centuries range from Altichiero da Zevio who founded the Veronese art movement around 1375 with frescoes in Sant'Anastasia church (on whom I based my dissertation for my BA at UEA four decades ago), to International Gothic portraitist

Antonio Pisanello and perspectival disciplinarian Andrea Mantegna (see the San Zeno altarpiece of 1456 in Verona), and one of the gods of the High Renaissance, Paolo Veronese, whose frescoes and altarpieces of biblical subjects in Verona (1545) and Venice (1553) fuse religion with humanity. While carefully composed and executed, his pageants brim with emotion and spirituality. Not unlike the Evora, really! There's huge interest in the Lotus, a real head-turner in the land of the supercar. A romance of another kind is Juliet's balcony and courtyard where Romeo went a-wooing, according to the Bard, augmented by centuries of lovelorn graffiti.

As well as Vicenza, defined by its 20 palaces designed by Palladio, and Venice, La Serrenissima, the canal city 100km away on the Adriatic coast, the Veneto also boasts the city of Brescia. West of Lake Garda on the road to Milan, it's the start and finish of the legendary Mille Miglia road race in which a couple of Lotus Eleven Sports Racers participated in 1957, the year the original 1000-mile event met its demise through an excess of fatalities. Though neither Lotus finished, that of *Autosport* editor Gregor Grant acquitted itself well against the fancied OSCA opposition until the fuel tank ruptured. I've covered the Mille Miglia twice in recent years, and while the plethora of vintage European exotica is breathtaking, I didn't see a single Lotus. I'll leave that one hanging ...

Close to Brescia there's a new race circuit, Autodromo di Castrezzato Franciacorta, and we catch Stefano d'Aste and a few of his clients testing their Exiges and a 2-Eleven ahead of the next round of the burgeoning Lotus Cup Italy series. The Lotuses flit in and out amongst the miscellaneous Porsches, BMWs and Ferrari GT4 contenders and are never headed on this compact track. As for the Evora, accelerate at 100mph at 4000rpm and the engine growls and kick-starts a new lease of life, hauling strongly in every gear, and even in sixth it's going hard. The brakes are sensational: apply the slightest pedal pressure and there's instantaneous rallentando. Whatever the circumstances, curve, sharp turn or flat-out blind, rough or smooth, the Evora rises to the occasion. The seats are firm, with backrest adjustment but no tilt adjustment to the squab. At first acquaintance, they make Porsche seats seem like armchairs, but at the end of a long day's driving you emerge from the Lotus fresh as a daisy. Preferring to concentrate on the drive, I eschewed the radio and navigation system, though when reversing, its projection of the rearward camera's view is a useful aid in a mid-engined car, as are the sensors in the bumper. These came in particularly handy when negotiating remoter Italian lanes that occasionally petered out, necessitating a turnaround – and not just a three-pointer.

To celebrate some of Lotus' grander achievements we travel to Monza, 150km from Verona in the north-eastern conurbation of Milan. The circuit is set in expansive parkland peopled by joggers and cyclists, and we brought the Evora into the hallowed paddock for a taste of the past. Steeped in history, the 100-year-old circuit's atmosphere is almost palpable despite the modernity stamped on the stands and pits by the FIA and the voracious needs of F1.

Team Lotus first raced here in 1958, running Graham Hill in a Type 16 and Cliff Allison in a Mk 12, finishing 6th and 7th respectively, no mean feat on such a high-speed circuit. By 1961, there were no less than eight Lotuses entered, mostly privateers in Type 18s, though Jim Clark was in a works Lotus 21. He survived a brush with title contender Wolfgang von Trips, though the German ace and 11 spectators were killed as his Ferrari rolled into the crowd. It would not be the last time that Lotus and Monza were linked by tragedy, losing both Jochen Rindt and Ronnie Peterson to crashes here in 1970 and 1978. On a positive note, Lotus' first Monza win came in 1963 with Clark aboard the Type 25, and his next best result was 3rd in '67 in the Type 49. Jochen Rindt was pipped at the post by Jackie Stewart in '69, but Emerson Fittipaldi wore the laurels in 1972 to clinch the World Championship with the JPS 72, placing 2nd to team-mate Ronnie Peterson in 1973. Ronnie won again in '74 in the JPS 72, and in 1977 Mario Andretti marked a Lotus revival with a win in the new ground-effect JPS Type 78. He

Sampling the historic vibes in the paddock at Monza, the scene of five Lotus Grand Prix victories – 1963, 1972, 1973, 1974 and 1977. (Courtesy Johnny Tipler)

Largest of the Italian lakes, Garda's picturesque promenade provides a playground for the Evora. (Courtesy Jason Parnell)

Mario Andretti gets reacquainted with the Gold Leaf Type 49, in which he took pole for his first Grand Prix at Watkins Glen in 1968. (Courtesy Antony Fraser)

repeated the exercise the following year, only to be penalised for allegedly jumping the start, but, sadly, his team-mate Ronnie, who'd also won Monza in a March 761 in '76, met his end in hospital (allegedly through a surgeon's blunder) in the wake of the infamous shunt after the race-starter's bungle. Mario clinched the '78 World Championship for Team Lotus, unaware that his friend was fatally injured. The night before the race they'd raced each other back to their hotel – the Villa d'Este at Como – in a pair of press Rolls-Royces, Colin Chapman a helpless passenger screaming at Ronnie to slow down.

Mario, meanwhile, had previous history with Monza, having had his first outing for Lotus in the Italian GP in 1968, when Colin Chapman, still grief-struck from the loss of Clark earlier in the year, famously announced that having Mario on board was like having Jimmy back in the team; he'd qualified 10th, ahead of team-mate Jackie Oliver but, in the event, Mario was disqualified, having jetted across the Atlantic to do a USAC race and returned to Monza outside the 24-hour time limit for being absent from the track between practice and the race. Team Lotus results at Monza never quite matched up during the 1980s, though there were high scores: Elio de Angelis came 4th in the Type 87 in 1981 and 5th in 1983 in the 94T. Starting his meteoric rise to fame Ayrton Senna was 3rd in 1985 in the black-and-gold 97T and 2nd in 1987 with the Camel-liveried 99T. A wealth of

Any port wine in a storm: passing through the Douro vineyards on the way to Porto for the Historic Grand Prix meeting, 2009. (Courtesy Antony Fraser)

endeavour, courage and commitment over three decades top-line racing.

Replete with Italian culture, high art and a generous slice of Lotus heritage, I point the Evora north through the Alps for the home run. Sixteen hours later, I'm back in Norfolk.

PORTO PARADE
A Portuguese road trip and an historic Grand Prix
When I announced I'd be covering mid-July's Porto Historic Grand Prix, ears at Hethel pricked up. The Portuguese Grand Prix was a big deal back in the 1950s and, throwing down the gauntlet for Team Lotus, John Surtees was pole-sitter at the 1960 Portuguese Grand Prix in a work's Type 18. Stirling Moss qualified 4th fastest, and the two other works Lotus 18s of Innes Ireland and Jim Clark were 7th and 8th on the grid. In the event. Clark came 3rd, Ireland 6th, Surtees retired, and Stirling was disqualified for turning his Lotus around after a spin ...

As in 1960, the modern revival has the seaside esplanades of the city's Boavista district resonating to the bark of race engines yet again. A big Lotus presence was expected at Porto (Oporto in old money) in 2009, and the company's Mediterranean Sales Manager Joris Rosmolen creatively visualised a way of demonstrating the new Evora to the enthusiastic Portuguese fraternity. A car named after one of their own cities, even if we Brits pronounce it differently. So, after discussions with sales bosses Glen Moir and Paul Bing, I was given the green light to drive a brand-new Evora to Porto. The deal was to deliver it to Manager André Castro Pinheiro at the Porto Lotus dealership, and then join the Lotus cavalcade on its parade lap of the Boavista circuit during a lull in the race programme. It would remain in André's showroom for a week, and then I would drive it back to the UK.

It was this run, more than the earlier Verona trip, that showed me just what a fantastic car the Evora is for accomplishing a long drive in swift comfort. There are three ways to make the journey by road from Britain to Portugal. Having a home on the banks of the mighty Rio Douro in the north of the country, my family and I went there often. There's the Channel Tunnel from Folkestone to Calais, which puts you on the French highway quickest and with minimum fuss; or the P&O and DFDS ferries from Dover to Calais, which are pretty civilised – you can even get a glass of fizz and a massage in the Club Class and Premier Lounges, but those two options pitch you into the 1160-mile (1865km), three-day run through France and Spain and down into Portugal. Or, there's the 24-hour ferry crossing from Portsmouth or Plymouth to Santander. That way you have merely to get to one UK port or the other, and drive the 400 miles (650km) from the northern Spanish port to northern Portugal. Having pals in south-west France to overnight with along the route makes the long drive south the more attractive proposition, though, given a smooth crossing through the Bay of Biscay, with dolphins and whales to wonder at, the long ferry passage is actually the more convenient. Cost-wise, there's not a lot in it, if you have to do a couple of hotels in France and Spain, against the cost of the day-and-a-night Brittany Ferries crossing. This time, I opt for the Portsmouth to Santander route. I pick up the pearlescent white Evora from Hethel, appropriately enough a left-hooker, and collect my friend Teresa Cherfas from Barnsbury, London N1. We load our bags into the boot, and speed down the M3 to Portsmouth, off on the quintessential road trip.

The overnight crossing is a relaxed affair – though I have also made the nautical passage in the depths of winter when the Bay of Biscay was in turmoil and brown paper bags were to the fore amongst many of the passengers. As it is, we enjoy the prolific buffet dinner and take a tern (turn?) round the deck; indeed, some enterprising seagulls do actually hitch a ride from GB to Spain. Closer to Spain, whales are sometimes seen, this being the Atlantic Ocean.

The entry into Santander harbour is uplifting; a glorious, broad expanse of blue water enclosed by sandbanks and distant cliffs, lighthouses and forts, with attractive bourgeois suburbs morphing into city centre blocks. First off the boat – that's the advantage of booking Priority Boarding – and no problems driving

Disembarked at Santander, the Evora is set for the six-hour, 650km run to Porto. Teresa Cherfas supplements the satnav with proper road maps. (Courtesy Johnny Tipler)

over the ramps. The Evora sat-nav's primed for Porto, and we hit the highway. There's a hundred miles of new Autovia, pacing through increasingly arid hills to take us away from the coast – though the trek along Spain's northern shores is a tempting alternative with charming fishing ports like Ribadesella and Llanes to explore. Instead, the Spanish motorway network aims us south towards Burgos and Madrid, and at Osorro la Mayor we swing west onto the arable plateau of Leon. After an hour-or-so's blast through terracotta flatlands, it's south to Benavente. I hog the outside lane as the inner one is badly pitted, despite its infancy. Along with the roadhouses here, where goat shanks exceed plate size, the hilltop fortress town of Puebla de Sanabria is the most convenient staging post en route. I reflect that it's here where my son, Alfie, fainted from the heat when we did the journey four-up en famille in a Porsche 911, back in the days when both he and his sister, Zöe, were small enough to go in the back. The Evora's rear seat would at least have kept the kids out of the fierce sunlight, which is partly what did for him in the Porsche as they were right under the back window. Now, though it's just as hot, and thankfully the Evora's air-con comes into play.

Another route option at Puebla de Sanabria is to skip the motorway and climb the high Sierra on a B-road, then drop down from the moors to Bragança in Portugal's top north-east corner. The new Autovia is so good heading west, though, that we elect to stick with it, turning off at Verin and heading down the A-road towards Chaves and the Portuguese frontier. It's another of those blink-and-you-miss-it border crossings; blockhouses and control booths long abandoned, as they nearly all are across mainland Europe. Apart from Switzerland where they want money from you.

Soon enough, it's time to stretch the Evora's legs again. These EU-funded motorways are absolutely fantastic – great swathes of rollercoaster tarmac and concrete snaking far into the distance; tunnels and viaducts galore making short work of hills and valleys. There's so little traffic that straight-lining the bends makes sweet work of it. The Evora tucks its nose in, the Pirellis bite and it tracks perfectly through the swerves without lifting the throttle. Cruising at 100mph is supremely effortless, plunging down huge gradients, powering the turn at the bottom and winding on the torque for the ensuing climb. This is one heck of a power trip. Coming off the motorway at Regua, the last few miles to my house are on twisting two-lane B-roads that hug the contours along the Douro valley, so high above the river that it's mostly out of sight till the final dive-down. We make it to our riverside palace some six hours from Santander, but who's counting when the drive is so sublime?

The following day I drive the Evora to Lotus Portugal's showroom on Rua Delfim Ferreira. On the way I'm singled out in a line of traffic by cops at Marco de Canaveses just because I'm in a sports car. Feigning a lack of lingo makes for a short interview and I'm soon satnav'd into Porto's northern suburbs. Cobbles and undulations notwithstanding, the Evora rides bumps with ease. A bevy of Lotus fans has already gathered for the run down to Boavista circuit, and they swarm all over the Evora. I know one or two club members already and am rapidly introduced to many more. Some have brought copies of my books for me to sign, which is always a buzz. After coffee the retinue of Lotus ancient and modern sets off for the seaside, the Evora bringing up the rear. I've handed it over to André, and I sit in with salesman Gustavo Rossi in a Type 121 Europa. A cut above the Elise in the comfort stakes it may be, but the civilised car I remembered from my Marrakech drive a year or two back is a far tighter squeeze than the Evora I've now become dangerously used to.

Ensconced aboard Ramiro Santos Silva's old school Elan S4 for the show-off parade lap, I have a 360º view from the cockpit for taking snaps. There's plenty of Lotus interest in the actual races, from Types 23 and 47 to Seven and Cortina, and I have the pleasure of Noel Stanbury's company, the '70s JPS promotions fount and Team Lotus Commercial Manager in the '80s and early '90s, enjoying being back in the fast lane for the

The Evora parked up at Tipler's residence at Arêgos beside the Rio Douro, northern Portugal. (Courtesy Teresa Cherfas)

The Evora aroused much interest among the cognoscenti while on display in the showroom of Lotus' Porto concessionaire. (Courtesy Johnny Tipler)

The Evora joined a line-up of Lotus, classic and contemporary, for the parade laps of the street circuit during the Porto Historic Grand Prix. (Courtesy Johnny Tipler)

Co-driver and valet Teresa cleans the Evora's bug-splattered front-end during a refuelling stop at León in north-west Spain. (Courtesy Johnny Tipler)

weekend, team-managing John Fenning's ex-Gilles Villeneuve Ferrari F1 car.

After a sublime week in the Douro, it's time to collect the Evora and head back to the UK, and it's largely a case of making the outward journey in reverse, except accomplished at a higher velocity. It's one thing meandering when it doesn't matter what time you arrive at your destination, but when there's a designated ferry to catch you can't afford to hang about. An early start, then, and empty roads enable a 5-hour 20-minute journey from Arègos to Santander, including a breakfast stop and a refill near León. Constant vigilance is the name of the game at these speeds. While the satnav announces imminent Gatso cameras, Teresa and I ceaselessly scan bridges and verges a mile ahead for lurking radar. A pair of Spanish motorcycle cops that hove into view ahead of us means throwing out the anchors and cruising at their steady 80mph till they turn off.

Quayside Santander, another amusing spectacle. While the Evora's front-end is seriously bug-blattered, it's no way as hard-core as the Exige coincidentally in the next lane: at the end of a Mediterranean tour, its entire body – an indistinct white hue – is stained pink with wine from a bawdy festival and evidently, while still wet, liberally peppered with sand, so it's camouflaged like a Pink Panther desert Land Rover!

It's an uneventful crossing, though this time we're dead last off the sumptuous Brittany Ferries Pont Aven ship at Portsmouth, priority boarding notwithstanding. It's a hassle-free run into central London, where we're celebs for the day under the curious gaze of pinstriped commuters. Then I'm off again for the short haul back to Norfolk. What a marvellous car this is. It's got everything: looks, panache, performance, stowage, and the all-important feel-good factor.

The copper-brown Evora with gold coachlines, ski-racks and skis, recalls 007's Esprit Turbo from the 1981 movie For Your Eyes Only. (Courtesy Jason Parnell)

Heading for Geneva – in a blizzard. (Courtesy Johnny Tipler)

DOWNHILL RACER
Slip-sliding away? Not in an Evora on winter tyres

Reprising 1981's James Bond epic *For Your Eyes Only* aboard a ski-racked Evora, I went off-piste in the Alps. I'd gone to Geneva, gateway to the Alps, luxuriating in the Hotel D'Angleterre beside the lakeside promenade, on a Lotus Cars press trip.

I've driven an Evora down from Hethel, crossing the channel Dover-to-Calais on board a DFDS ferry, then following the Autoroute payage, passing Reims and Dijon in the small hours. Lotus' Club International marketing exec Caroline Parker was in a blue Evora, mine's the copper-brown car with gold coachlines that's been got up at the factory to recall 007's Esprit Turbo from the 1981 movie *For Your Eyes Only*, complete with a pair of ski-racks and skis. It's a cool idea for a retrospective promo video, linking classic Lotus with new. That Esprit in the Bond film was the one with skis on the top and lattice-spoke wheels, currently resident in the James Bond Museum at Keswick, complete with Sheena Easton soundtrack, in which 007 AKA Roger Moore dodges ski- and motocross bike-mounted baddies bent on taking him out over a disputed nuclear submarine tracking device.

No such dramas await us. Or maybe they do: the Lotus Club International trio – Caroline, snapper Jason in the Galaxy pursuit car, and me, in the 007 car, have been mysteriously routed to Geneva via a twisting mountain pass, thanks to a unanimous decision on the part of three independent satnavs. So what? Well, there's a blizzard going on, that's what. And while the Evoras, with their Yokohama winter tyres, cope surprisingly easily, the Galaxy hire car has regular all-year-round rubber, and Jason is in trouble, ice-sliding every which way on the snowy downhills, and with little traction on the ascents. We decide it's not worth risking his camera equipment in a car wreck, so we buy a pair of 'chausettes' from a roadside garage, literally socks that pull over and straddle the Galaxy's front tyre-treads in the manner of snow chains. Thus shod, Jason can make the slippery descents and, soon enough, we're cosily ensconced in Geneva's lakeview Hotel D'Angleterre, surrounded by a select coterie of media folk, here for an exclusive press reception.

My plan for next day is to is to head round the southern, French, side of Lac Léman – Lake Geneva to us – and track up into the Alps on a circular route that will bring me back to the hotel in time for the glühwein. One slight problem: it's snowing hard. Looking out across the lake at breakfast-time through the dining room's picture windows, big waves are breaking on the opposite quay half-a-mile away as yacht masts sway in the gale. A few lonely water taxis ply, but rather them than me. For the trip, the tinkle of tiny hooves and sleigh bells spring to mind, rather than the earthy growl of the Evora's Toyota six-pack.

Right now, it feels good to be in the warm bosom of the hotel. Bond would suave it out for sure with a mid-morning Martini. But that's not how we Lotus-eaters operate, and as the weather clears and the sun burns through, it's off to work.

My co-driver is broadcast journalist Jenny Forsyth. She's an ardent slipper and slider, so at least one of us will be all right if the white stuff prevails. We pinpoint Morzine ski resort as the focus of our run. While Morzine is in France's Haute-Savoie department and the Rhône-Alpes, Geneva – Genève – is in Switzerland. Just. The map shows a neat round trip that takes us out of Geneva on the meandering lakeside road towards Évian-les-Bains. These are roads for pottering along, which the Evora does well enough; evaluating its real prowess is something for faster A-class roads and of course they're affected to a greater or lesser extent by snow. A prod on the throttle to encourage the back to twitch is instantly controllable with a touch of opposite lock, and it's so much fun that restraint is called for.

First port of call is the unspoiled medieval village of Yvoire. Rambling stone houses, pantile roofs, shuttered windows, some festooned with evergreen creeper, famed for their floral arrangements in season. There's a turreted castle by the lake shore, plus café, hotel and an onion-dome church spire. The village is on a promontory and delineates the 'grand lac' from the tail-end 'petit lac.' Geneva is at the western end of the banana-shaped lake – the largest body of natural freshwater by volume and area in Europe – and Montreux, of jazz festival fame, is at the other. There's a view across the water to Lausanne on the Swiss north shore, reachable by ferry. This is bathing territory, or probably would be if it were summer. Past Thonon-les-Bains and its giant hedgehog roundabout is Évian-les-Bains, where the bottled water comes from. The stones at the lake edge are covered in ice and there's no aquatic activity here just now, though the town comes alive in summer when the waterborne spree and the belle époque casino come alive. For now, the colourful street market stretching up the hillside will have to do: "Saucisse! Fromage! Venez goûter," shout the traders. Try before you buy!

The origin of Évian's natural spring proves to be the pagoda-shaped La Source Cachet, where the beneficial waters gush with alarming velocity. And here you can actually replenish your empties for free. In the town centre is L'Éspace Thermal, where sufferers of gout and digestive ailments seek Le Cure, endorsed by the French Ministry of Health. The waters are also harnessed commercially at the clinical Spa where, if it takes your fancy, you can be subjected to aquatic massage via powerful jets and a dip in pure Évian. You could probably bathe in asses' milk for the same price.

On snow tyres, the Evora coped admirably with the snow-covered roads heading for Morzine, where the skis would come into their own. (Courtesy Jason Parnell)

From Évian, a winding lane flanked by four-feet high snow banks takes us south, going ever higher, though I feel totally secure with the Evora's traction on the difficult surface – power overlaying ice. But it's still a bit of a relief to join the salted D902 that follows the cascading course of the River Dranse de Morzine, running between conifers heavy with snow and rock walls festooned with icicles and frozen waterfalls. Every building, whether house, shop, church or hotel, is timber-built, plain pine, in a concoction of low-pitch roof, shuttered windows, with elaborately coloured, carved balcony and log-store beneath. One thing they all have in common is their own personal ski slope on top. A couple of feet of snow is marvellous insulation.

For Jenny, Morzine is nirvana, and for me, devout Ski Sunday armchair enthusiast, the lure of 80mph in the raw, without any protective bodywork, let alone a steering wheel, is crazy but thrilling. Morzine is at the heart of one of the most extensive ski regions in the Alps, known as Les Portes du Soleil. It's one of the bigger resorts, and there's a good vibe here with lots of families in action on the runs, with more than 400 miles of tended trails and slopes, unlimited off-piste terrain and 207 lifts. Half the slopes are beginner or intermediate level, and today's fresh snow means deep powder so it's not so fast. Apart from one of the best snowboard parks in the French Alps, and that means big jumps, there's a range of slopes for all levels of skiing skills and experience, with several firms busy offering tuition. You can also do skidoos, sledging, snow treks, cross country skiing, and paragliding – a couple of which come close to landing on our cheese-and-onion tarts as we snack outside the Action Sports Café just below the ski-lifts on Avenue de Joux Plane.

There's plenty of snow at the top of the first lifts, and higher up they say it's perfect. The Winter Olympics are just getting under way on Whistler's Cypress Mountain and I'm anxious to impress

It was easier finding grip on the powdered snow than following the compressed icy tyre tracks. (Courtesy Jason Parnell)

any roving talent scouts looking to recruit last minute subs for the British team, say, as a sweeper in the curling competition ... But Jenny can't resist the call of the slopes, so, since we have the kit on board, we offload the Evora's James Bond skis and she heads for the six-seater chairlift; there's a new high-speed one and a magic carpet too, so there's no hanging about. You can do this at Morzine (2277m top) and higher up at adjacent treeless Avoriaz (2460m top) reachable from Morzine by cable car, gondola and high-speed chairlift – arrive and slide, just like Bond does in *For Your Eyes Only* along with delectable Olympic skater Bibi Dahl (say it quickly) – provided you've got waterproofs and boots, just turn up and ski. 'Ah, Mr Bond! We've been expecting you.' Every other shop seems to sell or rent ski wear and those in between are patisseries or café-bars, like La Buddha Bar and Café Chaud near the Pleney lift. We round off the Morzine experience at the après-ski Dixie Bar with a Jägerbomb: Jägermeister meets Red Bull.

Leaving Morzine, we rejoin the D902 at Les Gets and track off to the spectacular D907 at Taninges, winding through pristine meadows, woods and more chalet villages, the Evora tucking its nose in obligingly through the corners, so competent I feel I could do anything with it. Not long after Taninges I spot a huge Ibex mountain goat climbing the steep roadside, sporting the biggest pair of horns you ever saw, and I slam on the anchors in order to snap his portrait. Uh-oh. He turns and heads back down the hill. Poor hungry chap, I think, and toss him an uneaten baguette sandwich. A bit later I reflect that this is his gig – he's an opportunist, a roadside rustler. No wonder he looked so well fed!

We arrive back on the outskirts of Geneva at Annemasse. The urban border crossing into Switzerland is unmanned and the satnav routes us through the suburbs and into downtown Geneva, where every watchmaker you ever heard of has their name emblazoned on belle époque building fronts – including 007's favourite, Omega. Around the end of the petit lac there's the homecoming façade of the Hotel D'Angleterre. It's been a really great day, a fabulous drive in a brilliant car and a few healthy slides on the white stuff to boot. I congratulate Miss Jenny Penny on her downhill prowess. "Flattery will get you nowhere, but don't stop trying," she says. Pure Moneypenny.

So far so good. The following day, we check out Mont Blanc, Chamonix and charming ski resorts like Argentière, the location Jason chooses for his photoshoot. We're less than an hour south-

Winter wonderland: typical snowscape near Morzine, Haute-Savoie. (Courtesy Johnny Tipler)

Letterboxes wear snow bonnets in Morzine. (Courtesy Johnny Tipler)

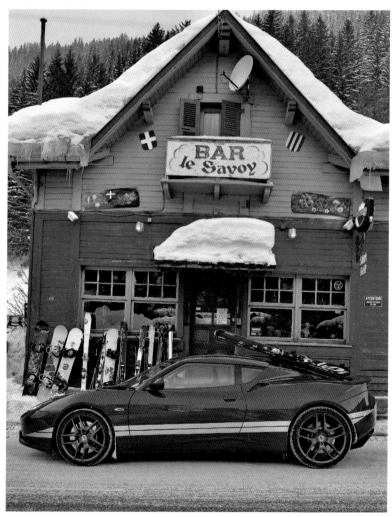

Pausing for a hot chocolate and a shot at a popular café in Argentière. (Courtesy Jason Parnell)

east of Geneva, and the views are breathtakingly spectacular against a blue sky, so jagged and pointed and, simply, high. The hot chocolate's pretty good too.

But trouble awaits 007. The hotel concierge has been tasked with hand-washing the Lotuses each day to remove the inevitable coating of salt and grime, and for a couple of days that works just fine. Then a new guy comes on duty, and he's either not briefed properly or prefers taking a short cut, and so he runs the Bond Evora through a carwash. Big mistake! The flails make short work of the neat ski-rack installation, ripping it off and damaging the car's engine cover. So, Lotus has to fly out Kieran who created the racks specially for the Evora so he can mend it. Our Bond car is to be involved in a major promo action the following week, involving 007-style stunt skiing over at Chamonix, so that, and a fashion show in Geneva also based around the car, hang precariously in the balance. Happily, in a matter of hours it's fixed, and Kieran's on his way home to Hethel. Meanwhile, we have another appointment, 250km south near Briançon, at Serre Chevalier ice racing stadium.

DANCING ON ICE
Evoras tackle an extreme freeze-up

Check me out: I'm wearing more studs than the punks at a Clash gig. To be precise, 384 studs on each Pirelli Sottozero. Even so, my Evora is broadside on the ice at Serre Chevalier Stadium.

This snow-lined labyrinth lies in the French Alps between Grenoble and Briançon on the outskirts of a ski resort of the same name, on the N91 to Alpe-d'Huez and Grenoble via the Col du Lautaret, often prone to a winter white out, and which management pledge to keep open "no matter what." Serre Chevalier is where the Andros Trophy kicked off in 1990, an instant hit that grew into a seven-round series, where big name famous race and rally drivers like Alain Prost and Olivier Panis drift sideways round the frozen corners. That's for the quasi-modernists, but in 2008 Serre Chevalier hosted a stage of the Monte Carlo Rallye Historique, which saw 300 or so classics pirouetting on the compacted snow. It's a series of hairpins connected by three 100-metre straights, with a cone or two to define entry, apex and exit points in the turns – Mickey Mouse, in racing parlance.

Serre Chevalier Stadium is a race circuit defined by vast snow banks and cones to aim at on training days. (Courtesy Johnny Tipler)

Matt Becker fits studded tyres onto an Evora ahead of some fun on Serre Chevalier's ice racing stadium, Briançon. (Courtesy Jason Parnell)

Ready to tackle the frozen curves of the Serre Chevalier stadium. (Courtesy Jason Parnell)

We brought four Evoras down from Geneva to evaluate on the ice rink. First job is to swap the winter tyres for studded ones, and to that end we've brought sets of slave wheels for each car. Ride and Handling maestro Matt Becker checks pressures. They're all front wheel size as there's no benefit to having broader rear boots in these circumstances. As used by World Rally Championship contenders, the Pirelli Sottozero Snow tyre has similar dimensions to the Evora's normal P-Zero tarmac tyre, though it's a more flexible carcass and narrower tread pattern, which increases pressure on the ground and therefore grip. The ice clench is delivered by those 384 studs that protrude by 1mm – the Swedish version goes to 1.5mm. And boy, what a transformation they make. While the Yokohama winter tyres grip admirably as I drive up the back lane to the Serre Chevalier workshop to make the wheel swap, once on the studded rubber the car becomes an indomitable beast, as sure footed as a polar bear on an ice flow. Julien, one of the resident instructors, takes the wheel to show me the lines, braking and acceleration points, when to turn and how much lock to apply. He's smooth with the controls even though this is his first acquaintance with an Evora, and he's plenty impressed with the power delivery. As we get faster I note his techniques, including that perverse Scandinavian flick when at the approach to the turn the car is jinxed the opposite way to the desired direction of travel to unsettle it before turning in at the last second, with oversteer corrected by varying degrees of opposite lock, and all the while trail braking late into the turn and applying accelerator roughly two-thirds the way round.

Then it's my turn. Not too much throttle to get away, then overdo it on the straight to feel the back-end break away, partly to see what traction is available and also just for the joy of it. I try to emulate Julien, and I get the braking about right but I keep giving it too much lock so it understeers into the deeper snow on the outside of the corners. "You must look where you want to go," he says; "keep looking around the bend, like a motorcyclist!" And that is helpful. Don't just look at the apex and especially not the snow bank on the outside of the turn, but peer as far ahead as possible, applying throttle to get the traction going out of the bend and switching from lock to lock to catch any fishtailing. I'm hauled into line by Julien: "Don't use too much accelerator," he says. "Wait till you're more than halfway through the turn." I drift sideways a time or two, but I manage not to spin. You can just mosey round letting the studs do the work, but what would be the point of that – this is a Lotus we talking about after all! After half-a-dozen hectic laps of

Balletic sequence: setting up the Evora for the hairpin turn ... (Courtesy Jason Parnell)

Power on, opposite lock applied .... (Courtesy Jason Parnell)

... straighten up and accelerate ... (Courtesy Jason Parnell)

hard concentration I'm stir crazy and have to have a break from what's a steep but addictive learning curve for an ice virgin like me. Look out you Ice Road Truckers, here I come!

There is more to ice driving than studded Pirellis and a few well-placed cones. The car's suspension setup also has a vital role to play. As Matt Becker tells it, "The suspension setup on a low-grip surface is more important than people think. That's because the way the car behaves when it's got less load on the tyres and suspension is more evident when you're on a low-grip surface which shines a harsh spotlight on any suspension inefficiencies. The Evora is comparatively easy to drive around this circuit because its behaviour is very benign, very progressive." Matt has certainly put the hours in to get it to that state, working with his colleagues at Lotus and Bosch to refine the suspension during the Evora's gestation, as we've seen, testing it relentlessly at Bosch's lakeside facility in the frozen north of Sweden at Arjeplog. The componentry is top-notch, he says: "the Eibach springs and anti-roll bars and the Bilstein dampers are all set up well and that really is apparent in these sorts of conditions. Eibach products are to a very tight tolerance in that they're repeatable on every Evora, so that there's no change in the behaviour pattern between different vehicles."

As a driving experience, Serre Chevalier Ice Stadium is similar to Arjeplog, "Except," qualifies Matt, "that here we're using studded tyres. Even though in winter in Sweden you'd probably run studded tyres in normal motoring, we don't run studs on the Bosch test track because we're replicating adverse road conditions with the car in the format it would normally be used in. So, on the ice there we only have winter tyres fitted, which are significantly better than summer tyres but not as grippy as the studs. In Sweden the suspension is something you've got to consider when setting up the car: you don't want to make it too stiff when you're using it on a snowy surface where you've not got much load pushed into the tyre. So, on snow it needs to be soft and supple – if it's too stiff you won't be able to achieve the level of grip you need with a winter tyre. You won't get enough downforce because you're not applying so much load, and that's because there's a cushion of snow between the tyre and the road surface. Ice is a very difficult surface to drive on, so again, you don't want the car to be too stiff. If the tyre hits a lump of ice you'll see a significant change in its behaviour, so it's important that the car is fluid over all surfaces. Snow and ice are more significant considerations than people think."

So, what does Matt make of Serre Chevalier? "It shows the car off well and it's fun to drive around," he says. "It's pretty much ice with a layer of snow on the top, so it's a manufactured surface. That is different to our testing grounds in Sweden,

Matt Becker guns the Evora along the snow roads beside Serre Chevalier Ice Racing Stadium. (Courtesy Jason Parnell)

One of our Evoras takes a break from dancing on ice at Serre Chevalier. (Courtesy Jason Parnell)

which, apart from the frozen lake, is more like typical snow on country roads. That's why you need studded tyres here; snow tyres would struggle to adhere on the polished surface." Punked up with studs, the Evora is more graceful ballet than flailing pogo, though eminently capable of holding its own in any ice-rink mosh-pit.

There is a postscript, and it's not a good one for our lensman, Jason. He's up on the snow bank that lines the track, photographing the Evoras circulating the stadium. One second he's upright, the next he's on the deck writhing in agony. He's slipped off the bank and fallen badly while trying to save his camera. The fire brigade and an ambulance arrive to cart him off to the local fracture clinic – they know about such things only too well in these parts. Turns out he's got three fractures, one in his leg, two in his ankle. The snapper snapped. Plastered up, Lotus will fly him home the following day.

For the rest of us, meanwhile, there's the long haul back to Geneva through the interminable Fréjus Tunnel between Modane and Bardonècchia – 8 miles (13km), 10 minutes and €43 long. I travel with Matt Becker, and more gloomy tunnels follow, where the Evora's V6 bark echoing off the barrel-vaulted ceilings provides consolation. The transition from dark to light on exit is blinding. The fantastic hairpin road up from Briançon to Montgenèvra is made for the Evora, and we scream up the helter-skelter, powering hard out of each turn, and then later on, straight-lining the sweeping curves at speed on the Autoroute. At around 100mph the suspension firms up and the car takes on a much more purposeful attitude, as if to say, 'OK, let's get on with it then.' In the process we manage to hit three countries in the day – Switzerland, France and Italy. Apart from in the tunnel, the outside temperature never climbs above zero, even in the city. The roads are salty and grimy, and fortunately we've topped up the Evora's screenwash reservoir with undiluted fluid, otherwise the jets would be frozen up. We savour the prospect of that beer and the resident soul trio back at the Hotel D'Angleterre's cosy Leopard bar.

BLUE OYSTER CULT
A tour of the Evora's East Anglian homeland
Lotus' Suffolk backyard brims with attractions, from windswept coastal castles to mellow medieval mansions, though some specialities give pause for thought.

"Tilt your head back and swallow," says the waitress. Gulp! The object of her instruction is a plate piled high with native Butley oysters – freshly harvested and resting in their crusty, crinkly shells, along with sliced lemon, naturally. The Lotus Club International team has paused to sample the wares of Orford, one of Suffolk's most southerly coastal villages, while on a

An Evora road trip on the Suffolk lanes crossing Westleton Heath. (Courtesy Jason Parnell)

Early morning beside the Blyth river estuary and its jetties near Southwold. (Courtesy Jason Parnell)

countywide progress with a new Evora. Perhaps nothing sums up the sea as much as oysters – they even taste of it, and it would be a sin not to slip a few while on the doorstep.

The drive from the Hethel factory has taken me cross-country via Bungay and Halesworth to Southwold, an impressive Georgian coastal town famous for its renovated pier, jazz festival and, real-ale boozer's delight, Adnams Ales. From Blythburgh on the A12, the five miles of A1095 to Southwold has a couple of long straights and a selection of up- and downhill, on- and-off camber bends that make you think Targa Florio – except that here the asphalt is un-Sicilian perfect – and the Evora tucks its nose in and grips, cool and assured as a mamba on a mission. You just know on a road like this you are in the perfect driving machine. All too soon the 30s come up for St Felix's School and the game's over ... for the moment. Forking right in Southwold, the Blyth river estuary abruptly meets the North Sea a mile or so south of the town, where tide and current thrash one another in a maelstrom of white water – and woe betide errant bathers any time of year. While the fishermen's huts at Walberswick on the south bank have largely departed, those on the Southwold side flourish, with fresh fish-'n'-chip cafes and fishmongers doing a roaring trade – in high summer. Now, at 6.00am when I've arrived to catch the sun-up over the saltmarsh for our lensman's benefit, it's completely deserted as the watery, wintry sun casts a pink hue on the blue car's flanks.

One word describes Southwold in November: bracing. In

The black fishermen's huts alongside the Blyth estuary offer fresh seafood and café lunches in the summer season. (Courtesy Jason Parnell)

winter, the wind has direct access from Siberia, although today it's warm and wet from the west. Snaps in the can, I point the Evora south, sticking as closely as possible to the coast. I touch the main A12 for 100 metres at Henham, site of the fab Latitudes pop festival where the previous summer's eclectic gig saw Grace Jones, The Editors and Mika strut their stuff. Come to think of it, I played here once myself in a Samba band. On the right, Blythburgh church stands sentinel above the marsh – that's the one with Black Shuck's claw marks in the door.

A fast climb over the narrow but enticingly swift B1125, and I bear left onto brown-brackened Dunwich Heath. This gives me the option of fowl play at Minsmere bird sanctuary, though the tearooms at Dunwich are certainly the more attractive prospect for breakfast. No chance of posing the Evora by Dunwich church for photos, though, not unless it's endowed with 007's submersible Esprit's capabilities, because virtually all of Dunwich's medieval port (half the size of London in the 12th century) is at least 100 metres out to sea, ravaged to oblivion by merciless waves during centuries of coastal erosion. Legend has it you can still hear the church bells when wind and water allow it, but the closest I come is going like the clappers. Ding-dong!

The Evora's needle-sharp handling and smooth acceleration through the gears are a real pleasure on these undulating backroads, where its sure-footedness is confidence inspiring. It soaks up the bumps effortlessly and it tracks unerringly through the twists and turns. This is a Lotus playground and it's here that the Evora displays is agility, the slightest steering input is all that's necessary to take most corners fluently, using second and third gears and applying scant pedal pressure to get it going – which also contributes to admirable economy: when fill-up time eventually comes around I brim it from a predicted 50km to 550km for £50. Miserly.

In pouring rain, the brakes respond without demur and the single wiper clears the screen most efficiently. I skip between puddles and piles of horseshit, which aptly defines my rural aspect, a reminder that mounted riders, not to mention cyclists and tractors, could lurk around any corner. Deer are an omnipresent hazard too, prone to stand stock-still in the middle of the road when mesmerised by headlights.

At the southern end of dune-flanked Sole Bay is the giant white golf-ball of Sizewell nuclear power station. Love it or loathe it, Sizewell was the saviour of neighbouring Leiston, providing jobs when Garrett's time-served steam railway engine business evaporated in the 1960s through a failure to go diesel-electric. A century ago, what's now the former train factory's Longshop Museum in the middle of Leiston, housed one of the UK's first production lines.

It's a short hop to Thorpeness, a quirky Tudorbethan seaside village with a boating lake and golf course, its lofty landmark House-in-the-Clouds – an ex-water tower – visible for miles around. A stretch of dog-walk heath, and we arrive in august Aldeburgh, arts-and-crafts hotels and shops along its main drag offering a melange of merchandise – including one of the best fish-and-chip shops on the east coast. On the shingle beach I contemplate Maggi Hambling's giant stainless-steel scallop shell, then a bracing hike to the yacht club blows away the cobwebs.

The A1094 is another spur off the A12, East Anglia's easternmost conduit, and I zoom inland to Snape. Russet sails of the moored, gaff-rig sailing barges flag up the Maltings, where Benjamin Britten's Canticles is playing live that evening, along with choral music by Purcell, Schubert and Schumann. By day, scores of weekenders throng the complex of converted warehouses, sifting through antiques, furniture and art for souvenirs and Christmas baubles. Or are they just sheltering from the rain? A lightning flash briefly illuminates the Lotus: my very own Electra Glide in Blue.

There's buried treasure in these parts too. Eighty years ago, Sutton Hoo burial mounds opposite Woodbridge on the east bank of the River Deben yielded the largest haul of exotic Anglo-Saxon gold jewellery and armour ever found, the legacy of a number of 7th century warrior noblemen – possibly even King Raedwald – and cohorts, interred with their ships, a ceremony similar to burials in the contemporary Beowulf epic. It's still

Brick arcading at Snape Maltings, home to the Britten-Pears concert venue. (Courtesy Jason Parnell)

an archaeologist's paradise – as portrayed in the 2021 movie *The Dig* starring Carey Mulligan and Ralph Fiennes. Sutton Hoo is bleak this time of year, but a walk in the parkland conveys something of the flavour, and the exhibition in Tranmer House fills in the gory details.

With only agricultural traffic to worry about, it's a relaxed run along the B1078 to Orford. (Courtesy Jason Parnell)

Caution's called for, as these are tractor roads and there's mud on the turns, so it's a relaxed run along the B1078 country lane to Orford. I park in the square opposite the Butley Oysterage. It's what this little village is famous for – oysters, plus the added attraction of Richardson's Smokehouse right behind the Oysterage restaurant. Could be straight out of *The Archers*. The oysters are reared in Butley creek, down-river from Orford quay, where the Alde meets the Ore at Havergate Island, home to over-wintering avocets. Pinney's, who run the Oysterage Restaurant, have an oyster boutique down by the quayside, and the seascape has a powerful allure: a vast expanse of choppy grey water, sheltered from North Sea by Havergate Island and, further out, the long pencil-line spit of Orford Ness. Isolated markers dot the spit skyline: pylons that serve the BBC World Service transmitter; strange pagodas that dispensed nuclear energy till 1971, and the distant lighthouse that dates back to 1627. There's also the largest of England's anti-Napoleon Martello towers, paradoxically rentable for hardy holidays from the Landmark Trust, its squat barrel walls a counterpoint to Orford's soaring medieval castle. Crab boats and dinghies bob in the swell, while quayside, the Lady Florence is moored, the 50ft ex-naval launch converted to floating restaurant and set fair for a three-hour on-board lunch- or dinnertime cruise up river to the RSPB nature reserve. It's captivating, but I must motor on. I'm billeted in picturesque Westleton at the Crown Inn, so I backtrack 20 miles. One moment it's a staggering sunset and I bask in its golden glow. The next, the heavens open and in an instant I'm a drowned rat.

The full gamut of English rural life is here: the chapel opposite the pub is full of secondhand books, traffic-stopping flocks of ducks have taken up permanent residence on the village green, and tractors trundle in for fettling at the old-timer garage.

Medieval and modern: the Evora by the ruins of Leiston Abbey. (Courtesy Jason Parnell)

There's a log fire crackling in the bar, and the Evora is soon the talking point among the groups of young-free-and-singles dining out.

I head west via Framlingham, pausing to check out its decayed castle. Suffolk's not flat by any means, positively hilly in these parts, exemplified by hidden dips where tarmac's scoured by over-zealous boy-racers bottoming out. Roadside hedges, ditches and woods abound, houses colour-washed Suffolk pink, orpiment yellow, more or less any hue you care to name. A diversion casts me deep into single-track lanes where corners are right-angles that mark the field boundaries, yet the Evora takes them in its stride with no dramas. Where one might be tentative in a 911 on some of these turns, the Evora provides the confidence to take them vigorously. Back on the high road, the A1120, I'm cruising at 2000rpm in sixth gear, giving 50mph – 80kph in this left-hooker, but the most rapid progress is made using third and fourth along roads like this.

Next up is Stowmarket, home to the Museum of East Anglian Life and the late DJ John Peel, who once explained to me in an interview that his elderly Mercedes was never molested, nor even locked up, because he made a habit of leaving apparently soiled underwear on the driver's seat. Fair play, John. Back in 1970 his radio show was called *Top Gear*, though he was a Morgan rather than a Lotus man.

My Suffolk tour has been both fascinating and heart-warming, sampling aspects of the county one might simply by-pass on a major road and not even be aware of. And now, at least this Evora is run in and ready to take its turn on the European press fleet.

Making swift progress across Dunwich Heath, keeping a wary eye out for deer crossing the road. (Courtesy Jason Parnell)

Elisa Artioli and the Evora

Granddaughter of Romano Artioli who owned Lotus from 1993 to 1998, Elisa was just three-years old when the Elise was launched. It's a matter of genius that Artioli named the new sports car after his granddaughter, bequeathing her an original Series 1 Elise in the process. Having passed her test, Elisa began driving her heirloom Elise, leading Lotus expeditions on road trips such as the Stelvio pass, not far from her home in north-east Italy. Having interviewed her in the past about her Elise, I quizzed her in 2021 about the Evora, and here's her response.

"The Evora is the first all-new Lotus after the Elise. To me, it is incomprehensible that the Evora never got the attention and appreciation she really deserved. She is a true Lotus in every respect. I am mainly used to driving my Elise, which is why it took me a while to understand the Evora. To start with, she felt very large. One thing I still find challenging is the lack of a rear view, especially when parking without using the back-up camera."

"After spending some time driving the Evora, getting used to her and adapting to her different dimensions, I felt that she has all the ingredients of a proper Lotus. The crisp response to the steering, preciseness and responsiveness is something you only find in a Lotus. The Evora is so powerful and agile. What I especially like is the sound she makes, and the fact that she has a lot of luggage space. She is a very good driving companion on long distance trips. She is definitely a perfect Gran Turismo, but also a powerful supercar. That, however, didn't stop me from taking her to Ikea as well!"

"This particular model is my favourite one. It is a model-year 2017 Evora 410 Sport, featuring the ducktail spoiler and slats rather than a window in the back. Also, it has many carbon-fibre parts like the roof, the front splitter, the mirrors, the seats and the front access panel. The combination of dark grey with the yellow accents and the black forged wheels just looks awesome."

As we went to press, Elisa had just ordered the very last Elise and Exige to be built, both finished in gold with golden upholstery!

The Elise's namesake, Elisa Artioli, also fell in love with the Evora – in this case a GT410 Sport. (Courtesy Joerg Uhr)

Elisa Artioli describes the Evora as "a perfect Gran Turismo." (Courtesy Joerg Uhr)

SIMPLY GORGEOUS
The magnificent V6 echoes off the soaring cliffs that form Cheddar Gorge

It's a bonus when a road trip enables me to drop in and visit friends or relatives along the way. My eldest son Jules lives in Bristol with his wife, Dr Vic and their two kids Florence and Henry (name checks all round!), and after weekending with them I drove the Evora GT410 Sport press car the short distance to Cheddar Gorge for a rendezvous with my snapping chum Antony Fraser. I know this actual car quite well, having had a blast in it previously to visit lifelong Lotus aficionado Malcolm Ricketts at his Hertfordshire base – when my local pal Steph Ewen took some pictures of it.

Approaching Cheddar Gorge from its southern start point, you pass at 20mph through the bustling village, easing quickly into the winding rock corridor. The utterly dramatic chasm provides an awesome setting for a circumspect blast along its serpentine thoroughfare, a steady incline (or syncline, as you'll want to go up and down it) designated the B3135, demanding total attentiveness due to the tendency of the feral goat population to suddenly dart – or leisurely amble – across the road, plus reckless cyclists (wrecked cyclists pending) hurtling downhill and puffing up it, as well as striding hikers armed with Scandi-poles. Those soaring cliffs, sometimes sheer on the south side, are staggeringly awesome – 137m at the max – but beware falling rocks: the Evora received a near miss when a tumbling stone ended up beneath the car.

Nothing comes as challenging and purposeful as a GT410 Sport. (Courtesy Stephanie Ewen)

Time traveller: posed between a couple of its elder siblings – Malcolm Ricketts' Elan and Elite – the Evora GT410 Sport looks a big car. (Courtesy Stephanie Ewen)

King rocker: the Evora did have a near miss in a Cheddar Gorge car park. (Courtesy Antony Fraser)

Soaring cliffs, flocked by wild goats, provide a craggy backdrop for an outing with the GT410 Sport. (Courtesy Johnny Tipler)

Snaking through Cheddar Gorge in the Evora GT410 Sport was a highly pleasurable experience. (Courtesy Antony Fraser)

Longleat Estate owns the southern flanks, and the National Trust the northern ones. A Longleat employee drew up at one of the generously broad parking areas in his 90TDi, bent on collecting the garbage he said was ejected nightly by nocturnal party types, though the large quantities of rubber deposited by their doughnuts did prove impossible to harvest. Why are some of us so slovenly? Meanwhile, birds of prey including buzzards, kites and falcons circle high overhead. The limestone canyon was formed 1.2-million years ago when water trapped below ground and frozen in the caverns melted after the ice age, and subsequently the erosion by the underground Yeo river, causing slumping that's visible in the in the bare rock strata, and creating cave formations known as Gough's and Cox's holes. It would be remiss not to endorse the local cheese too, available from a number of outlets in the eponymous village at the foot of the chasm. After the shoot, as Antony, Mrs Fraser and I tucked into very good pizzas, we observed comedian Susan Calman doing a cheesy promo to camera.

That's sorted out the environmental circumstances; now, about this wonderful car.

Behind the wheel of the Evora 410GT Sport
I bend my body into the cabin of the Evora 410GT Sport. The perfect driving position is found in a trice. The Recaros are very comfortable, though trimmer Karl advises me they are outsourced – whereas the dashboard and everything else in the cabin is trimmed by him and his staff at the factory. The Alcantara headliner and door lining looks and feels nice. The Recaros, then, are supportive, providing just the right angles of posture in relation to the steering wheel, gearlever and pedals. From the driving seat, all the instrumentation and gauges are orientated ahead of the driver. The rev-counter and speedo are housed directly in front, with tyre pressure sensors and fuel gauge to the left of the rev-counter, temperature and time and mileage to the right of that, and above the centre console is the Sat-Nav – doubling as rear camera screen – and hi-fi monitors, with the various heating and ventilation controls on the console ahead of the gearshift. Looking in the mirror, through the back

Masses of cosseting Alcantara with flashes of red highlights characterise the 2021 GT410 Sport's cabin. (Courtesy Stephanie Ewen)

Helming the GT410 Sport through the twists and turns of Cheddar Gorge. Cheesy grin? Naturally. (Courtesy Antony Fraser)

window I can see the throttle working on the engine, which is the same as the Exige. As for the slats of the rear window, which might be irritating, you simply disregard them, so you're juggling the view out back between the slats. In the side mirrors you're seeing the cooling ducts responsible for feeding air into the engine bay, and then ahead, the sculpted lines of the front wing architecture curving sexily towards the front of the car as the nose drops away in front. From the driving seat you can't actually see the nose, but you can judge where it is from the curvature of the front wings.

Before you fire the car up, you have to remember to blip the immobiliser on the key fob, which is fair enough, though slightly irritating and awkward on a modern supercar. To start, turn the ignition on, press the start button to the right of the instrument binnacle, and it suddenly bellows into life in a rather ostentatious way. Accelerator, clutch, brakes and steering are all spot on and a treat to use. I'm finding the six-speed shift a bit notchy, not

Panning shot of the Evora GT410 Sport speeding up the Cheddar Gorge incline. (Courtesy Antony Fraser)

uncertain, but there's a need to make a deliberate shift between fourth and fifth, let's say, or third and third, rather than pushing the lever vaguely across the gate in those plains. Having driven a few miles, the cabin's quite warm by now, but the ventilation is not brilliant. The windows are still misted up even having had the blower on, and I have got the stylish air vents open.

On the move, and when it's thoroughly warmed up and everything is working properly, when the revs are around 6000rpm, the closest comparison I can think of is one of those David Attenborough programmes featuring a bellowing elephant … quite an awesome sound. Around the twists and turns of my Mendip Hills course, it tracks absolutely unerringly, and I can adjust minutely the turn-in angles just by lift-off understeer or foot-down oversteer. You think it round the corners, looking as far ahead up the road as possible in the same way as you would if you were riding a motorcycle, especially when going quite quickly through the bends. There's just as much feedback in this car as there is with the more raw Exige, and it's everyday-usable with that very definite track-orientated focus. In general terms, the Evora – and the two other models in the Lotus line-up – contrives to be harder-edged than, say, a standard Porsche: it's firmer, more engrossed, and at high speed it's completely focussed. The Evora is alive from the moment you start it up, but where it really comes alive and obliges you to pay close attention is over the ton, just holding its line absolutely beautifully through curved sections. You have to be in the right gear at the right time so, let's say, going into and coming out of a roundabout, second gear is the optimum one to be in. A glance at the rev counter reading 3000rpm in top gear means we're doing 90mph, which is a sort of a bench mark.

The subtle nuances of right foot pressure on the throttle pedal do mean that one can finesse the nose as to which way the car turns in or out. On a tighter corner, the more you put your foot down, the harder it grips. The Evora has always had an excellent ride, so whilst the suspension is very firm, you really don't feel kind of pot holes where the road is in a bad shape. Ride and handling are quite as well sorted as any Grand Touring car, and that brings me around again to the point of a GT, which is to be able to accommodate two – maybe two-plus-two infants – in high style, plus their weekend luggage for a grand tour of La Cote D'Azur, or some-such. In this respect the Evora is barely credible: sure, the main seats are really great and the cabin is easily spacious enough. But, and it's a relatively small butt, there's no front luggage boot like Porsches have, so you are limited to the rear cargo space, which is not a lot bigger than the Elise, and arguably only adequate for a couple's weekend rather than a fortnight's road trip. Sure, there is space in the rear cabin for storage, unless you happen to have kids on board. Having said that, my grown-up (as in adult) daughter Zoë has travelled side-saddle in the rear of an Evora on a cross-county culinary mission without expressing too much discomfort.

Swishing this way and that, the GT410 Sport stuck like glue through the twisty bits. (Courtesy Antony Fraser)

There's an element of the masochist about motoring across continents four-up in a two-plus-two, though. When we had a house on the Rio Douro in northern Portugal I drove my Porsche 911s several times via France and also on the overnight Brittany Ferries crossing to Santander, and Zoë and Alfie did suffer in the heat of the car's rear passenger compartment as we crossed the sun-drenched northern Spanish plateau. Whatever, I could drive an Evora all day every day and be totally on cloud nine. As for Cheddar Gorge, like any honey trap, it's fine to do now and again, certainly once, perhaps as a bookmark on a journey. Next stop, Glastonbury ...

THE HILLS ARE ALIVE
With the sound of Evora ...
Lotus press officer Alastair Florance made available an Evora GT410 on two or three occasions, and it's an interesting comparison with the GT410 Sport, reviewed on my Somerset sojourn, in so far as it is less fulsomely specified in terms of the quality of cabin upholstery and trim, and comes over as not quite as well appointed or luxurious as the GT410 Sport. It does contain more sound-deadening material than its more focussed sibling, allegedly in a bid to improve day-to-day usability, but I contest that as a valid premise, because you don't buy a Lotus to be mollycoddled. It also endowed with a marginally softer ride – milder, less aggressive, maybe – though I would take either model across the continent and they are pretty close in the ride department. The softer suspension is said to have been inspired by the settings offered on standard American-market Evoras, and it's fitted with Michelin Pilot Sport 4S tyres rather than the GT410 Sport's Pilot Sport Cup 2s: having run these on a 911, I can see why the Pilot Sport 4Ss are the less challenging tyre, especially in the wet. Though, given these Pirelli Sport Pilot Cups on the GT410 Sport, I only felt it twitch once in the wet with power on.

Apart from external cosmetic enhancements on the GT410 Sport, what mostly separates the two siblings boil down to cabin equipment: the GT410 gains a few items from the options list as standard equipment, including cruise control, rearward camera activated when reverse is engaged, air-con, and an information-and-entertainment system with Apple CarPlay. The doors boast armrests and the Sparco sports seats can be heated.

The GT410's powertrain is the same as that of the GT410 Sport, which is the most recent development of the supercharged 3.5-litre Toyota V6, producing (an eponymous) 410bhp and 295lb/ft of torque. The six-speed manual transmission is standard issue, with six-speed paddle-shift automatic available as an option, although just the manual has a Torsen limited-slip diff. It has a valved exhaust system like that of the GT410 Sport,

The Evora GT410 certainly has dramatic appeal. (Courtesy Alex Denham)

The flowing lines of the Evora GT410's bodywork are stunning from every angle. (Courtesy Johnny Tipler)

Driving the Evora GT410 through a Lincolnshire forest. (Courtesy Alex Denham)

moderating the volume of the V6 most of the time, until the sound suddenly opens up at 4500rpm and above when going for it. It has the familiar noise-button allowing the valves to be open throughout the rev range, and although they can be permanently closed, engine breathing is better when they're open. The loud control certainly promotes a sense of speed and images of on-track bravado – there's that old boy racer again – and it is possible to imagine that throttle response does seem sharper with the valves permanently open. Acceleration and performance are identical, whichever transmission is used, both variants getting to 60mph from zero in 4.1-seconds, though the manual has the higher top speed, maxing out at 186mph to the auto's 174mph. The GT410 is equipped with AP Racing brakes and like all Lotuses, there are no qualms as to its slowing and stopping capabilities.

With a low scuttle and relatively high seating position it

provides great forward visibility; given the Evora's rear quarters, the view of the road behind is always going to be slightly compromised, though door mirrors and internal mirror provide adequate imagery. Comfort-wise, the Sport has it beat, though of course, when navigating the road, considerations like this become insignificant. The controls are also where you would expect them to be: the washer, the lights even the starter button, all switches fall neatly to hand. There's a decent sort of resistance to the gearshift.

On the move, the way the power comes in is absolutely spontaneous, and it's certainly rapid in a straight line. The steering matches the throttle response, in that it's totally direct and positive, providing clarity and feedback, and turning in or out with the slightest input. In the corners it's so wonderfully controllable, while the ride is good on everything except a really pot-holed road when you feel every bonk. It's fun to play with the Sport mode, which you feel immediately firms the suspension up. But in practice the noise-volume is fun once or twice but in the normal course of things it's not really worth bothering with. Though it does make me smile!

I've arranged to meet my snapping pal Alex Denham at Cadwell Park circuit in the Lincolnshire Wolds, and she's beaten me to it, sat outside the main gate having discovered it's all locked up due to "the pandemic." The drive across west Norfolk and the Fens is unrewarding in respect of roads and scenery, and it's not until I reach Horncastle that I'm suddenly plunged into the undulating Wolds. Here, there are examples of every kind of bend, gradient and cambered corners, swoops up and down, blind summits and troughs, all adding up to some really marvellous driving roads. Which the Evora takes in its stride – laps up, if you like.

As I've always averred, what's marvellous about the Evora is the directness and spontaneity, the immediacy of feel, its tactile nature and contact with the road, and the responsiveness of its controls. I spot a snake-like section of road across a shallow valley, half a mile ahead, and I'm already mentally setting the car up to optimise the blacktop, gauging which gears I'll need – second and third – how much throttle to deploy, allied to over- and understeer, helming it round on the wheel and balancing the accelerator pedal. It's kind of seat-of-the-pants thrilling when you know you've got it just right, aided and abetted by the finest GT chassis out there.

From market town Louth, Alex and I motor over to the Lincolnshire coast just south of Mablethorpe to the optimistically named Miami Beach. On a different Lotus gig, to Daytona Speedway, no less, this is kind of like Daytona Beach without the buildings. We can get the car virtually onto the sand for Alex to

Beside the seashore, the GT410 takes in the Lincolnshire coastline at Horncastle. (Courtesy Alex Denham)

Out of the woods? The GT410 branches out into a spot of forestry. (Courtesy Alex Denham)

take some shots. After that, we drive back inland to a forested glade where more photography takes place. A cuppa outside a transport café, and we're done for the day.

By the time I'm back in Norfolk it's dark. Here's where we go negative. On dipped beam the Evora's lights are just not sufficient to penetrate the night, and they're only just about good enough on main beam, to the extent that I keep hunting in vain

Towering spectacle: driving the Evora GT410 is a religious experience. Almost. (Courtesy Stephanie Ewen)

for another switch for some non-existent auxiliary spotlights. I think, on the whole, they're just not up to scratch, regrettably, even comparing unfavourably with the lights on my 20-year-old Boxster.

And, talking of Porsches, the GT410 is stickered at £82,900, making it £100 more expensive than a standard 992 Carrera. The 911 might have a higher output in terms of brake horsepower than the Lotus, and some would argue superior build quality, but it's the way that the Lotus deploys its performance that's the main difference. The Lotus provides a more engaging driving

Working at the car wash, yeah! Giving the GT410 a shine ahead of a photoshoot. (Courtesy Alex Denham)

Lotus Cars' public relations supremo Alastair Florance declared that the Exige 430 Cup would be the fastest car I'd ever driven, point to point, and he wasn't wrong. (Courtesy Laura Hampton)

experience than modern Porsches. It's a luscious dancer rather than an amiable pugilist. Of course, you can hype the handling and performance of virtually any Porsche sports car, whereas a Lotus is your loaded weapon straight out of the box, armed and ready to go. It is poised and responds instantaneously, intuitively, to a greater extent than the Porsche. In art history classes, the mantra was all about 'identify, date and compare,' and that is what we're up to here: the Porsche feels a rather more solidly created, crafted machine, and I suppose that, being made of metal rather than with fibreglass panelling, it's almost bound to be the case. They are very different characters.

The supercharged GT410 power delivery is in the shove-you-in-the-back category, albeit subtly and not alarmingly dispensed. It's a visceral experience: while other high-performance and high bhp-rated cars I've driven provide a similar sensation – take the Porsche 991 Turbo – the Lotus is an acute experience to the Porsche's more muted one. I've also driven Ruf's RT12 and many other offerings from the maestro of turbocharging, and also 9ff's 1000bhp 991-based GTronic, both immensely powerful, though refined and not in a crazy way. The Lotus reminds me more of a racing car – and actually I have to mention another Porsche in this instance, the 917 ("69 Le Mans car) I drove around Donington Park circuit in 2019. On second thoughts, I'd also cite the Exige 430Cup, another Lotus press car I borrowed in 2020, of which Alastair Florance foretold, "I bet you this is the fastest car you'll have ever driven, point to point." And he wasn't wrong.

MANX FOR THE MEMORIES
The Evora tackles the daunting Isle of Man TT course
Last throw of the dice: in late September 2021, I took an Evora GT410 to the Isle of Man. When I picked it up from Lotus' Hethel plant, PR Alastair Florance said that the final Evoras were just then going down the line. So, where to take my old friend HL66LCL? I'd never been to the Isle of Man before, and having

Waiting at the factory gate, having collected the Evora ahead of the run to the Isle of Man – which is for the riders as well as the drivers; the riders go a lot quicker, though. (Courtesy Zoë Tipler)

viewed Guy Martin's video *Closer to the Edge*, I was inspired to sample it for myself. On four wheels, rather than two ...

A lap of the TT course when the event isn't on is a hit-and-miss affair. On public roads, the course is subject to normal traffic conditions, ranging from tractors, lorries grinding slowly up hills, traffic lights at junctions such as St Ninian's corner and later on entering Ramsey, plus several delays at road works – getting shots at hump-back Ballaugh Bridge was impossible on account of road works and lane closures. And 20mph limits during schools coming out, such as in Kirk Michael. On one lap, the Mountain road was closed because of a traffic accident at the Bungalow – an establishment itself shut due to road improvements. But enough of these qualifiers to enjoying a good experience on the course; there are ample opportunities for appreciating the bends – and the straights – when you've got 37 miles and 220 corners to play with (actually 37.73 miles, or 60.72km). That said, so long as speed limits are adhered to in villages – get this – there is NO national speed limit on the open road. 100mph? How about 130? I can only stand in awe of TT racers who hit 200mph in places. They do 160mph through Crosby; we can only dawdle as it's still an urban area and a 30mph limit. Given the terrain and the omnipresent perils, these speeds seem nothing short of suicidal on two wheels; I'd be hard pressed to dare do them on four, even with the roads closed to public access. For the record, the current lap record is 16 minutes 42.77 seconds – an average speed of 137.45mph – set by Peter Hickman in 2018. Under the circumstances, timing our laps in the Evora would have been futile.

The obvious differences between being on a car and on a bike are the extreme vulnerability to hazards. Starting with the surface, it's decent blacktop for much of it, but some is godawful in other places. Plus, manhole covers, drains and the hotchpotch of bad repairs in urban areas like Douglas, Ramsey and Kirk Michael. If a car with the superlative compliance of the Evora feels the undulations, imagine what goes through the frame of a race bike and the rider's body. Simple things that we take for granted loom large at high speed, and demand total vigilance: whilst you can remember the main landmark corners like the Gooseneck, there are plenty of lesser bends that are unmemorable and they can catch you out, as can camber changes. There's the pitter-patter of all the undulations you're experiencing the whole time through the steering wheel, but on a bike, that'd be through your wrists, your arms and your torso as you're lying on the petrol tank. And when the weather changes abruptly, like it did on one lap where we got a few drops of rain, or bright sunlight, which can dazzle depending on time of day, creating pronounced shadows as you rush through avenues of trees, banks and buildings. Much of the course is delineated by kerbing, some quite high and some fairly innocuous, but the majority of the kerbstones are alternating black and white, lending a chequered flavour to the course. There are miles of implacable stone walls, mostly very well maintained, which constitute yet another hazard. I hate to think what would happen if a wild animal strayed onto the course – which must happen occasionally. And sure, the TT does have a reputation for fatalities and, sadly, over 250 riders have died since the first event in 1907. But let's not dwell on that. It's about the challenge of pitting machine against the landscape, going as fast as possible on a minuscule strip of ever-changing tarmac.

Here's how the landmarks unfold. The races start off on the outskirts of Douglas on the long straight of Glencrutchery Road, beneath the Race Control tower, the principal grandstand and the pits, where you'll see refuelling taking place and riders

LOTUS EVORA

On the Isle of Man TT pits and starting grid alongside Glencrutchery Road, the Evora GT410 is overlooked by the event's control tower. (Courtesy Johnny Tipler)

One of the first landmarks on the TT course as the route zooms out through Douglas' suburbs is Ago's Leap, a tribute to the Italian hero Giacomo Agostini getting air during his heyday between 1965 and 1972. His average speed most years was 104mph. (Courtesy Laura Drysdale)

Another landmark on the 37-mile course is the Ginger Hall Hotel, normally obscured by spectators during TT time. (Courtesy Johnny Tipler)

mounting up. It's an urban scenario till Union Bells, where there's the Railway Inn, standing on the outside of the course, which overlooks the right-hand curve as it falls downwards toward the left-hander in the village. Then there's Governor's Bridge, which is a slow right-hand hairpin, followed by a left-hand curve through the Governor's Dip, and then an uphill exit to the right. To the cognoscenti they're iconic names, like Signpost Corner, Bedstead Corner, and Crosby Straight, which is a flat-out, 180mph stretch through the village, or Highlander, where bikes can be nudging 200mph on the straight downhill section, passing the Highlander restaurant. The Hawthorne Pub is a good spectating place, as is Glen Helen Pub, and Barregarrow, which is reckoned to be quite scary. In fact, a great deal of the course is exactly that: simply, scary. Innocuous Kirk Michael, where they bolt through the village at around 180mph – in what's normally a 30mph limit. Ballaugh Bridge, where momentarily the bikes get airborne off the humpback bridge, where there's quick flick left and immediately right. Sulby Glen, where once again the fastest bikes are going 190mph. The speed is incessant and unremitting.

Heading down to Sulby Bridge, where they brake really hard to zap round the right-hander, barely noticing the actual bridge, and then the Ginger Hall pub is on the right-hand side, where

Overlooking Ramsey and the coast from up on the foothills of the Mountain section. (Courtesy Laura Drysdale)

we've got a bookmark photo: right now, there's just a delivery van parked outside, but during race week you'd hardly see the place for spectators. At seaside Ramsey the pace is slowed by sharp corners and a very uneven surface. And then it's a long, steep ascent to the Mountain section, a moorland pasture setting with astounding views glimpsing most of the Island. The first corner is Ramsey Hairpin, literally a tight uphill left-hander, probably the slowest point – on our lap, at any rate. Soon after comes the Gooseneck, a sharp right-hander, with great braking and cornering action. And then the long haul, going higher and higher to the Bungalow, straight-lining sweeping bends, with a tramline crossing, and long views in all directions. Then comes Brandywell, a 100mph left-hander, just beyond Hailwood's Heights, the highest point on the course. We're getting close to the end of the Mountain section now, with Keppel Gate and Kate's Cottage, where the road descends quite steeply down

LOTUS EVORA

Coming up to the Gooseneck: a tighter line into the apex would be preferable, but ... (Courtesy Laura Drysdale)

The sharpest turn on the TT course is Ramsey Hairpin, an uphill left-hander. While the riders can use the whole road, it's tricky to emulate them in a car in normal road conditions. (Courtesy Johnny Tipler)

The Creg, one of the most popular locations on the TT course, with adjacent stand and marshals' post. (Courtesy Johnny Tipler)

At the wheel of the GT410, I'm constantly looking as far ahead as possible – just like a motorcyclist, though at a fraction of their speed on the TT course! (Courtesy Laura Drysdale)

to the Creg-Ny-Baa pub-restaurant – where we had a pit stop of our own, a really nice lunch, and Mrs T made out with Guy Martin – or at least a handily-placed photo of him on the saloon bar wall.

As for the Evora, it turns in impeccably, going round corners that have no apparent apex or exit, and even when I'd overcommitted due to unfamiliarity with the circuit, easing the wheel further to achieve the required turn-in. To a great extent, it's a sensory thing: I simply 'think' the car through the bends, the brain on autopilot, analysing and implementing direction control, the work of an instant. Some corners are spellbinding, and you think, ah, if I could repeat that, how satisfying would it be. Like a motorcyclist, all the time I'm looking as far ahead as possible. The Evora is as responsive as one could wish, perfect turn-in, requiring gentle throttle pressure, powering into and out of turns; a section like Quarry Bends is sublime, flowing through the half-dozen successive serpentine curves at 80 or so (I wasn't looking at the speedo, but it felt like it). And the turn of speed along the straights – with blind crests and hidden dips – is phenomenal. Whilst I'm in top gear for the fast sections, optimum gears are third and fourth, with second necessary for the Ramsey Hairpin.

Time for some R&R. A visit to the Manx museum in Douglas revealed that TTs were run for cars in the 1920s and '30s, so actually we were not masquerading usurpers doing laps in our

The Evora GT410 running down the hill from Kate's Cottage into Creg-Ny-Baa, probably going half as fast as the TT race bikes. (Courtesy Laura Drysdale)

The downhill straight from Kate's Cottage to the right-hander at Creg-Ny-Baa is fast and daunting, whether on two or three wheels. On four, not so much. (Courtesy Rebecca Sayle)

Evora rather than on a bike. Which would have been possible, as bikes and kit like leathers and lids can be hired. Contemporary four-wheel motorsport also features the Manx National Rally and the Rally Isle of Man, dating from 1963 and a round of the British, Irish and European Rally Championships, with a winners' roster including greats like Roger Clark, Tony Pond, Ari Vatanen, Patrick Snijers, Russell Brookes, Jimmy and Colin McRae, Mark Higgins, and all their illustrious navigators. Rally buffs will say, 'Why not major on the rallies, then?' Because the Isle of Man is, for me at least, all about the TT. A local rally was set to take place during our visit, and as we pootled along in the Lotus, a steward's car pulled alongside and inquired if we were doing a recce for it. We had to disabuse them, explaining that we were in fact searching for a Neolithic monument ...

There are a couple of other circuits on the Island: Billown, near Castletown, which dates from the mid-"50s, and there's St Johns, a 15-mile loop incorporating roads in the north-west, and Clypse, which incorporates 10 miles of the Mountain course. However, no visit to the Island is complete without a scan of the TT and Motor Museum at Jurby in the north-west. This totally eclectic collection has been accumulated by Denis Cunningham over the past 30 years, and includes unusual coteries of vehicles such as the 'Flower Cars,' huge American cars converted into pick-up trucks and allegedly employed at Mafiosi funerals. Or how about a dozen rotary-engined cars and bikes? There's a large number of restored Humbers, and Fiat 124 coupes styled by Vignale, in many cases with a restored car juxtaposed with an unrestored example. When I asked Denis whether the decomposing E-Type coupé would be restored to match its pristine neighbour, he replied that, if it was, it would simply be a restored Jaguar rather than an original unmolested car. The astonishing coincidence for me was discovering a friend and sometime colleague, racer and journo Mark Hales in the building, in the process of delivering a Citroën Maserati to owner and curator Denis Cunningham. We agreed that the last time we'd seen each other was, if not at the Goodwood Revival, then at the Spa Six Hours perhaps three or four years previously. We took a wander through the vehicles on display, and among the exhibits that interested us most were the rotary- (Wankel-) engined motorcycles. Which brings us back to two wheels: next time, I'll head over on a bike and catch the TT as it happens.

On the journey to and from Heysham to catch the IoM ferry, we hit the fuel crisis-inspired panic-buying week, and though we started off with a full tank I was minded to brim it a couple of times once the 'gauge' registered three quarters full. It helped not that the Yorkshire Dales pub I'd booked us in as a halfway house was shut – and had gone into liquidation some weeks prior. There was no such alarum at the Isle of Man's filling stations since they get their petrol direct from Ireland, and a full tank – via Hebden Bridge, the A1 and the wretched A17 – got us (slowly) all the way back to Norfolk. There was even a couple of hundred miles-worth left in the tank when I handed it back at Hethel.

The Point of Ayre working lighthouse marks the northernmost tip of the Isle of Man. (Courtesy Johnny Tipler)

CHAPTER 9
AFTERLIFE
Crowning the successor

The Evora's successor, the Type 131 Emira, was unveiled at Hethel on 6th July 2021. Speaking at its launch, Group Lotus Managing Director Matt Windle declared, "The Emira is a pure driving machine, and has a high level of driver engagement, it's also provided us with the opportunity to introduce new optional features, drive support systems and solution pack, and carries on-board technology such as ABS emergency braking and cruise control, and it has sensors that can warn you of hazards when changing lanes, and also a rear traffic alert – the kinds of features that, after a long day at work, say, they are going to assist you. Maybe if you've been using the car at a track and enjoying its dynamic capabilities, as soon as you start the journey home, you switch on the driver assistance and it looks after you."

Matt Windle joined Lotus in 1998: "I've been here on and off since then," he said. "I've had the privilege of working throughout the company, both on the shop floor and in the exec suite, and now I'm really proud to be in charge. I've had a great career; one of the most interesting periods was a setup called Tesla, and that was a great experience. At Lotus, anybody that works here will tell you that it just gets under your skin, and the products are absolutely fantastic. We'd always wanted to do a more usable every-day car and Emira gives us that, it's part of our long-term plans, meaning increased volume, and it's just the start of where we want to go to as a company. We decided it would be great to celebrate the last hurrah of the petrol engine, and we thought we would achieve that with the Emira, and it will take us to a wider audience. We've spent £200m on facilities and the factory

An Evora GT430 joins a brand-new Emira at the launch of the Fittipaldi-Evija at Hethel, October 2022. (Courtesy Stephanie Ewen)

Jenson Button shakes down the Emira at Laguna Seca, California, 2021. (Courtesy Lotus Cars)

is transformed. It's effectively the rebirth of Hethel. During the last few years we've built new factories and introduced new facilities across the sites, we've installed new technology and doubled the resources in that time, and that's not just at Hethel, we also have a manufacturing site in Norwich where the chassis are made. The Emira is the result of those investments, and it's a game-changer for Lotus, it's a really significant milestone. We've managed to achieve a striking design, with breathtaking performance, best-in-class on-road handling, outstanding aerodynamics and unbridled driving experience. We feel we've achieved a genuine supercar in a sports car, and it really is a shift in terms of practicality, comfort, functionality and technology. We managed all this at a competitive price, and there are two engine options as well: you have the choice of 3.5-litre supercharged Toyota V6, which we've worked with for a long time, and it's a

An Emira on the assembly line revealing a similar construction to the Evora, with rear sub-section housing the powertrain butting up to the central cabin section. (Courtesy Johnny Tipler)

AFTERLIFE

Each day at noon during the 2021 Goodwood Festival of Speed, fireworks heralded the Lotus spectacular featuring Evija and Emira. (Courtesy Antony Fraser)

Presenters at the Emira launch were Helen Stanley, Andy Jaye, Russell Carr, Matt Windle and Jenson Button. (Courtesy Stephanie Ewen)

The Emira launch was dampened by a cloudburst but, undeterred, Gavan Kershaw demonstrated the JPS Type 72 F1 car in conditions more suited to a speedboat. (Courtesy Johnny Tipler)

The Emira assembly line is housed in a spacious new factory building. (Courtesy Johnny Tipler)

blistering, bulletproof engine; or, our first new sports car engine for many years is the AMG Mercedes-Benz four-cylinder turbo, which is a highly efficient big output engine and incredibly fast too. It's the most accomplished road car to come out of the gates at Hethel."

The launch of the Emira was an early evening do, held on the Hethel test track, with a mysterious cube the size of a small house surrounded on two sides by long, draped gazebos. The press and corporate guests gathered under these awnings, and celeb hosts Helen Stanley and Andy Jaye began their banter. The heavens opened and, to his credit Matt Windle toughed it

out: it was really pouring down! There was a Zoom conversation between the Lotus presenters and a few hundred Lotus owners whose screens were projected onto the cube, the reason why the unveiling couldn't easily be postponed: there was a schedule to be adhered to. The rain eased off, and a succession of Lotus icons were driven around, emitting huge plumes of water – especially Gavan Kershaw in the JPS Type 72. Eventually, the two Emiras concealed in the cube were driven out, and – enter Jenson Button to declare the new model officially out there.

After the affable reception within the factory, a small number of guests repaired to the factory to see the Emira

Photographer Steph Ewen admires an Emira, resplendent in its launch livery. (Courtesy Johnny Tipler)

Right: The ProFleet elevating platform is crucial to Emira assembly, enabling component fixtures at heights convenient to the operators. (Courtesy Johnny Tipler)

Far right: Front left-hand corner of an Emira reveals double wishbones, coil-over damper, steering arm and ventilated disc brake. (Courtesy Johnny Tipler)

production line. Whilst the designated Emira build zone was far more spacious than the Evora had enjoyed, there were distinct similarities between the two models. Starting with the extruded aluminium chassis, and the Toyota-based 3.5-litre V6 powerplant. The Emira also has double wishbone suspension with coil-over dampers front and rear. Each unit is assembled on a ProFleet plinth that is elevated by the operators in order to attach and install components at a convenient working height. Whilst some cars were to showroom finish, others sported promotional livery reminiscent of the bizarre camouflage applied to warships ...

Chief Designer Russell Carr

We've heard Russell wax lyrical about the Evora, and he had this to say on Emira too. "When you say supercar, there are really three elements in play here; first of all, there's the obvious reference to the Evija Hypercar that we launched two years ago, and that was always intended to be a design statement as well as our first fully produced electric sports car; secondly, to combine beautifully soft sculpture with technical details, the balance that prove form and function; and finally, achieving the proportions: it's about getting the car to look low, hunkered down, as well as something that invites you to drive it. At the front of the car we've got this very sharp, attacking section that looks like it's going to cleave through the air, and that's very important to express feeling when it's standing still. Also at the front are the very distinctive headlights, which give the car a unique signature when it's looming up behind you at night. You'll notice that the lines flow beautifully seamlessly into the side sculpture, and the way they play with light and shadow on that helps describe a car that is both athletic and exciting to drive.

"We've been locked away in the studio for the last two-and-a-half years, slaving over this car, obsessing about the details, eating and drinking this car – but what it's all about now is sharing it with all the fans, the owners, the media and our family as well, all those people who wondered why we were spending so many hours at work every evening. Well, now they can see it was worth it. We've got a whole range of cars to work on at the moment, but the nature of our business is completely about confidentiality, so I can't share any details about those at the moment.

"The aerodynamics provide that tangible link to the science behind the beauty of the car. Every feature, every shape on the car's bodywork has a part to play in terms of the way the air flows around it, the way it guides the airflow into certain locations to do certain jobs. The splitter that catches the air at the front stagnates the air and pumps the splitter to generate downforce to counteract some of the upper body lift that's generated by the car. The air blades in the corner intakes, those characteristic boomerang vents in the bonnet, they physically suck the airflow out of the cooling system by creating vortices creating low pressure that drags the air out to help the efficiency of the cooling system and direct the airflow down and around the sides of the car. Those vents down the sides of the car are positioned precisely to be aligned with the two counter-rotating vortices along the back of the car, causing low pressure to suck out the hot air from the engine. Everybody knows that the car has got to be beautiful, but it's also got to work, and it's getting that harmony and balance between what it looks like and how it performs that is key. The rear view of the car is very powerful, very distinctive, and there are some beautiful details. This is our last combustion-engined car, and we really want to emphasise that, so you look at the exhaust pipes. The rear lights are very distinctive and the car looks broad and muscular."

"The interior should help people connect with the vehicle, so behind the seats you can put a couple of bags if you're going on a trip, and you've got storage in the doors, twin storage in the centre console, somewhere to put your phone, somewhere to put your sun glasses, and you have the classic glovebox as well, so it really is something you can use every day.

Continued overleaf

The 2000bhp Evija electric hypercar on the Lotus Cars stand at the 2021 Goodwood Festival of Speed. (Courtesy Johnny Tipler)

The sensational Emira at Lotus Cars hospitality building during the evening launch on 6th July 2021. Its supercar looks are redolent of the McLaren P1. (Courtesy Stephanie Ewen)

It's got twin-tier console, heating and climate control, and all the main surfaces are either leather or textile or Alcantara. Many of the touchable points are made of metal. It's a cabin interior that feels extremely sporty, with a maturity to it, so if you want to drive the car with passion it doesn't ask anything of you; equally, if you just want to use it on a normal trip it's a comfortable ride.

"When you work with the single objective of producing the ultimate sports car, all of the features that we have on the Emira are there for a very specific reason – form following function – and that's what a Lotus is, it's all about a collection of things that as a designer I consider vital ingredients, and I think it's the combination of its dramatic exterior looks, combined with a very sporty but very luxurious and usable interior, means that it's a car that you can use for any occasion, on the road or on the track. It's the culmination of decades worth of experience to produce what is I believe the best sports car in the world."

The Emira set to make its debut run up the Hill at the 2021 Goodwood Festival of Speed. (Courtesy Antony Fraser)

Aerial view of a 2012 Evora, highlighting the gorgeous bodylines of the original Russell Carr and Steve Crijns design. (Courtesy Johnny Tipler)

The Emira design process

When the Emira was announced, Senior Designer Daniel Durrant had been at Lotus for 12 years. During that time, he'd worked on the Exige LF1 Special Edition in 2014, the Evora 400 exterior programme in 2015, and in 2017, the Exige 430Cup. He had this to say about his involvement with Emira: "Overseeing design and development of the Emira exterior, I was thrilled to have my theme selected for the Emira. I have worked on many Lotus programmes in the studio, so to take the lead exterior design role was an amazing opportunity – and a huge responsibility." His first sketches were drawn in 2018.

His reference points included the Evija of 2018. "Lotus has a

Cabin furnishing

Interior Design Manager Jon Statham joined Lotus in 1997. Career highlights included input into the S1 Exige in 2000, the Esprit facelift in 2002 on which he was Lead Exterior Designer, a similar role on the 2-Eleven in 2006, and in 2016, the Exige 430 and Evora 430. His role on the Emira was overseeing the design and development of the interior. "To be involved in the full development programme, from initial sketches to production, has been challenging and rewarding in equal measure. The Emira is a big step forward on many levels. We've had a massive push on quality, technology, functionality, usability and desirability. It has a contemporary interior with good proportions; it isn't too hardcore, and it will have broad appeal whilst still being one 'For the drivers.'"

The small team working on the Emira cabin drew some inspiration from the Evija, "but there are also elements of the S1 Esprit in there," Jon admitted. "Jennifer Andriamamonjy's interior theme was the one chosen, and blends sportiness with modernity and quality. Harvey Rabenjamina worked on steering-wheel and seat design. Lotus created its first-ever digital design team to develop the HMI, and we've worked closely with them." Jon also credited digital modellers and studio engineer Josh Router. Work started on the Emira in November 2018, and it was clear that packaging the interior would be difficult. "We wanted outstanding ergonomics to make it as engaging as possible, and to achieve that while finding space for all the components was the challenge. I'm very proud of the cohesive nature of the interior design, which exudes quality. The instrumentation graphics are spot on, and perfectly suit the character of the car."

Getting a taste of what the future might hold ... (Courtesy Stephanie Ewen)

rich technical and visual library to draw from," he acknowledged. "I like the shapes of military fighter jets. There's often a softness to their overall surface forms, but with taught creases. Shapes found in nature can also be a great inspiration, like a shark nose or the muscular haunches of a cheetah, for example. To sum up the Emira exterior, it is sculptural, athletic, agile, elegant, and alive. The sculpture around the body-side air-intake is my favourite section. The surfaces are three-dimensional, and designed to channel the airflow into the body-side duct. There aren't many cars with this amount of form and drama in

Enthusiasts admire the Emira and Evija on show at the Goodwood Festival of Speed. (Courtesy Antony Fraser)

them: technical as well as beautiful. We always intended the design language to be that of a baby supercar. From a visual perspective, the biggest challenge was finding the right balance of sportiness and sophistication. It needed to look light, focused and agile, without ever looking too aggressive or intimidating. It also needed to look premium without being too conservative. From a technical perspective, the sensor positions and ADAS radar module were the trickiest parts to incorporate. The front-end is extremely low to the ground, just as a Lotus should be, but this creates challenges when it comes to positioning them. These components are very small but devilishly awkward!"

The styling was carried out in-house, following the normal Lotus design process that we've already looked at with the Evora, progressing from theme sketches, renderings, scale models and,

An Emira poised on the startline at the 2021 Goodwood Festival of Speed. (Courtesy Antony Fraser)

The clay buck has all the accoutrements of a complete car, including wheels and lights. (Courtesy Lotus Cars)

eventually, to a full-size clay model. "We tried many design ideas through the process before the final theme was chosen," Daniel recalled, "and the technical and aerodynamic package evolved along the way as we honed the design. We always wanted close visual and philosophical harmony with the Evija. It was important that both products looked related to each other, but also performed technically in their own way given they're very different cars. Overall, it's been rewarding and, as with anything that's worth doing, challenging at times. Getting to the point where the car goes into production is very satisfying."

Jenson Button shakes down the Emira
Who better to evaluate the brand new Emira V6 on the Hethel test track than 2008 F1 World Champion Jenson Button? "Every time I drive a Lotus I love that mechanical grip that it gives you, but now you wouldn't have that in high-speed corners they've worked really hard with the aerodynamics on the Emira and the aero is doing most of the work. Lotus engineers have worked really hard on the Emira to produce a lot of downforce for such a small car, which gives it a lot of grip through these high-speed corners, and that gives you confidence to push the car a little bit harder, which puts a big smile on your face. The Lotus is super-light, so you don't need 800bhp. You expect to have the mechanical grip on a Lotus, but to have the aero as well is something else. Everything on this car has a purpose: all these lines, the intakes, the curvatures, they have a reason for being there, and that's what I love most about it. It's light, it's nimble, and it's comfortable. I applaud everyone at Lotus; they can be very proud of what they've achieved with the Emira; it's a fantastic piece of kit. And you're doing it in such a luxurious place as well; this interior is so plush. A lot of sports cars are fantastic to drive, but you don't want to do long distances in them. But in the Emira I'm so comfy. The entertainment system is easy to work so you can get back to hammering through the gears. They've placed the gearstick in this car quite high, and it feels really racy, so from the steering wheel to the gearstick it's such a short movement which is great, it feels much racier and I obviously like that."

"I've loved Lotus cars since before I was legally allowed to drive one," Jenson affirmed. "When I was 17, I started off in something that was very far from a Lotus, but I drove past the Bristol showroom in Earl's Court, and there was a Lotus Elise

Jenson Button takes a turn around the Hethel test track in the Exige-based Radford Type 62-2, liveried in JPS black and gold. (Courtesy Lotus Cars)

outside, so I went in and I said, 'Can I drive that?' and they said, 'No.' One word: that's all they gave me, 'No!' 'Well, can I sit in it?' and they said, 'Okay, you can sit in it,' so I sat in the Elise, and I remember it just being very sparse: basically, it had a seat, three pedals, a steering wheel and a gearstick, but I just loved that feeling of a Lotus, when you got in it, you felt part of it, and that was what I loved the most. Obviously, the interior of these is very different, and it's moved on a long way. But I had the bug then, and I got to drive my friend's Esprit, which was very powerful and very exciting. And also in 2000, I went to Hethel when I was working with another manufacturer, doing laps and passenger rides, and at the end of the day I said, 'Can I drive the Lotus 340R?' and they said, 'Yes, of course you can,' and I jumped in it, and it was worlds beyond what I was driving around in at the time. So, yes, I'm very passionate about Lotus, and it's great to be part of the team and having the chance to drive these really amazing cars."

Jenson is also associated with Radford, founded by Harold Radford in 1966 to create bespoke luxury Minis – the Mini De Ville, as driven by all four Beatles and rock stars like Mark Bolan. Radford is reprising the Lotus Type 62, a Group 6 prototype raced in 1969 by John Miles, Brian Muir and Jackie Oliver. "They built two for racing in the '60s, and with my mates at Radford we're bringing back the Type 62, which I'm really looking forward to, something very exciting for the future, and we're very lucky to have been able to work with Lotus engineers on the platform and the development of it."

The Type 62 was a bit of an ugly duckling in its day, bedecked with add-on spoilers and flippers. While the Europa-derived Type 47 had been a Group 4 Sports Racer with 30 of the required 50 units built, the Type 62 that appeared in 1969 was a Group 6 prototype. It was intended to be a test vehicle for the new Lotus Type 907 engines that were earmarked for the forthcoming Elite road car. Designed by Martin Waide and presented in Gold Leaf Team Lotus colours, the Type 62 was a much bulkier car than the 47, with embryo nose fins and rear spoilers, and its bulbous wheelarches housed F1-size wheels and tyres – 12in rims front and 15in rear. The 1995cc four-cylinder twin-cam sat amidships in a multi-tubular spaceframe chassis, and it had a 16-valve cylinder head, fed by Tecalemit-Jackson fuel-injection, and developed 220bhp at 5000rpm. Transmission consisted of a ZF 5DS2 five-speed gearbox and final drive unit, and suspension

Endorsed by Jenson Button, the Radford 62-2 uses an Exige platform and running gear, and aims to emulate the Type 62, the 1969 test-bed for the Lotus Type 907 engine. (Courtesy Jules Tipler)

was by double wishbones, coil springs and damper units up front, and reversed lower wishbones, single upper links, twin radius rods, coil springs and damper units at the back. Its 12in ventilated disc brakes were mounted outboard front and rear.

In the 1969 BOAC 1000 at Brands Hatch, Miles/Muir drove the prototype to 13th place overall and won the 2.0-litre Prototype category. Two works cars subsequently raced in national and minor international races with Miles placing 3rd in the Tourist Trophy and 4th in the Trophy of the Dunes at Zandvoort. Significantly, the stresses of racing showed up flaws in the Vauxhall-derived engine block, but Lotus was able to address the problem independently, so the Type 62s had served their purpose. One was pensioned off to jazz bandleader Chris Barber who already had a 47, and the other was raced by Dave Brodie in 1970. Back in the day, we regarded it as possibly the ugliest racing car out there, given its protuberances, compared with a beauty like the contemporary Ford P68/F3L. Suffice to say that Radford's Evora-proportioned 3.5-litre representation of 2021, with its Exige underpinnings and Toyota V6 engine, is a way more elegant machine than its antecedent. Expect to see it in JPS Black and Gold as well as Gold Leaf livery. So that's another reference to a historic Lotus to bolster the marque's heritage.

Meanwhile, the Type 135 EV Elise replacement is also in development, due for launch in 2026. Matt Becker, for one, is sceptical: "The Emira is the Evora replacement, but the Elise replacement is allegedly going to be electric. I'm still not there with the EVs. I get that they're paying lip service to zero-emissions, which is not possible; it's not zero-emissions. Starting with power stations, the infrastructure's just not there to do it either. That's why I still think, for the time being, it'll be a blend of EV, hydrogen, synthetic fuels – as currently being developed by Porsche – and hybridization. You can't just say it's going to be EV. Because it's not."

EVs? It crossed my mind that, in the fullness of time, one might conceivably convert an Evora to electric: people are doing that with all manner of classic cars to keep them on the road. For now, though, we have a perfectly good fossil-fuelled model to play with.

A Type 124 GT4; artwork specially created for the book by Sonja Verducci.

Also from Veloce Publishing –

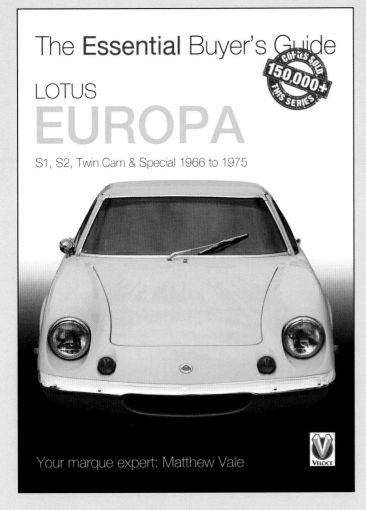

ISBN: 978-1-787112-87-2
Paperback • 19.5x13.9cm • 64 pages • 90 pictures

ISBN: 978-1-787112-86-5
Paperback • 19.5x13.9cm • 64 pages • 103 pictures

If you are interested in buying a Lotus Elan or Lotus Europa, theses books will give you the background information and technical details to ensure you purchase the car you want. Written by an author with experience of restoring classic Lotuses, the book will give you the knowledge you need to identify and assess any potential purchase.

For more information and price details, visit our website at www.veloce.co.uk • email: info@veloce.co.uk
• Tel: +44(0)1305 260068

Also from Veloce Publishing –

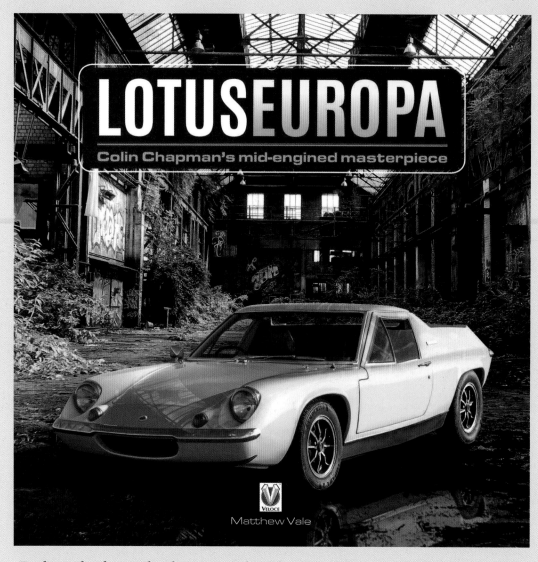

Explores the design development and production of the Lotus Europa, Lotus's first mid-engined road car. It covers the Renault-powered Series 1 and 2 cars, the Lotus Twin-Cam-engined versions, and the Type 47 racing models.

ISBN: 978-1-787112-84-1
Hardback • 25x25cm • 160 pages • 175 colour and b&w pictures

For more information and price details, visit our website at www.veloce.co.uk • email: info@veloce.co.uk
• Tel: +44(0)1305 260068

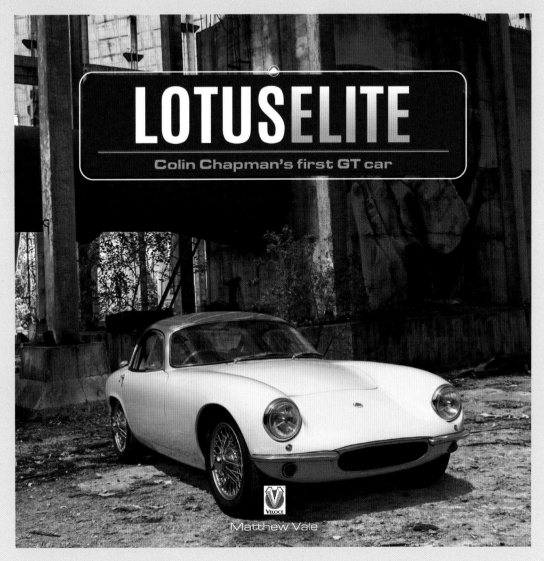

A comprehensive history of an iconic British GT car, this book looks at the Elite's place in Lotus' past. Interviews with ex-factory staff and current owners define the car's place in the classic car scene of today. Illustrated with colour photos and period advertising material, providing essential reading for the Lotus Elite enthusiast.

ISBN: 978-1-787115-17-0
Hardback • 25x25cm • 176 pages • 251 pictures

For more information and price details, visit our website at www.veloce.co.uk • email: info@veloce.co.uk • Tel: +44(0)1305 260068

Also from Veloce Publishing –

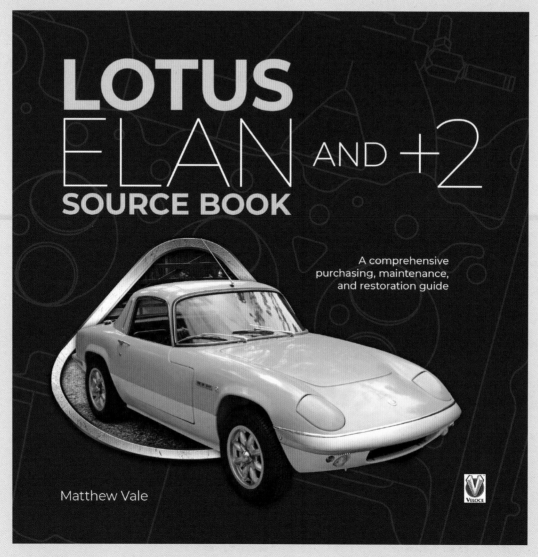

This book gives the Lotus Elan and +2 owner and enthusiast all the information they need to buy, maintain and restore one of these iconic British sports cars.

ISBN: 978-1-787114-59-3
Hardback • 25x25cm • 208 pages • 300 pictures

For more information and price details, visit our website at www.veloce.co.uk • email: info@veloce.co.uk
• Tel: +44(0)1305 260068

INDEX

9ff GTronic 216

Abbeville Circuit 77, 141
Agostini, Giacomo 218
Aida opera 182, 183
Alcon brake calipers 112
Aldridge, Geoff 91
Alfa Romeo 8C 10
Allison, Cliff 184
Alpine-Renault A110 10
American Le Mans Series (ALMS) 111
Andriamamonjy, Jennifer 229
Andretti, Mario 184, 185
AP Driveline Technologies 100
AP Racing brakes 130, 212
Apple CarPlay 211
Arêgos, Douro 188
Argentière 190, 194
Arjeplog, Lapland 14, 70, 96, 104, 105, 130-135, 198
Aroso, Pedro 189
Artioli, Elisa 4, 205
Artioli, Romano 205
Ashby, Mark 94, 98-101
Assembly process 165-172
Aston Martin DBX, DB9 Vanquish 6, 10, 28, 54, 84, 91, 111, 141, 143, 159, 173
Atlas Mountains 70
Audi
 quattro 58
 R8 84
 TT 19, 84
Autocar magazine 22
Autoliv seatbelts 75

Autosport magazine 184
Autosport Show 108

Bahar, Dany 8, 22, 23, 36, 137
Baillie, Michael 4
Barber, Chris 233
BARC 150
Bardolino 181
Barr, Matt 104
Bayley, Stephen 4, 54-61
Becker, Kristie 4, 30, 140, 141
Becker, Matt , 6, 7, 22, 23, 25, 30, 36, 51, 73, 107, 123-129, 137, 141-143, 196, 198, 199, 233
Becker, Roger 4, 6, 14-16, 23, 29-31, 36, 48, 91, 116, 141-143
Belgian Pavé 69, 73, 76
Bennett, Luke 29, 31, 33, 77
Bentley 17, 24, 134, 173
Bertone 40
Besenzi, Carlo 92
Bilstein dampers 92, 125, 198
Bing, Paul 186
Birch, Paul 94, 95, 97-102, 104, 107
Birchall, Clive 100
Blundell, Mark 122
BMW 17, 66
 328 153
 M3 111
 Z4 84
Böhme, Martin 75
Bolwell Nagari 93
Bond Bug 27
Bond films, James
 For Your Eyes Only 190-192, 195

The Spy Who Loved Me 30, 116, 117, 201
Bosch ABS 11, 14, 70, 73, 96, 102-105, 126, 130-135, 198
Böss, Alex 73, 130-132, 134, 135
Bradley, Derek 128
Brands Hatch circuit 42, 233
Brigham, Ian 128
Bristol cars 17, 27, 231
Brittany Ferries 186, 189, 211
Brodie, David 233
Bushell, Anthony 36, 38, 47-49, 51, 63
Button, Jenson 10, 224, 225, 231-233
Byblos Art Hotel 180

Cadwell Park Circuit 213
Campbell, Naomi 117
Car magazine 58
Carr, Russell 4, 11, 16, 22, 33, 36, 39, 40, 47, 48, 53, 61, 225, 227, 228
Carrera Panamericana, La 12, 152
Castrezzato Franciacorta circuit 184
Caterham Seven Superlight 128
Chamberlain Engineering 91
Chapman, Arthur 151, 152
Chapman, Clive 4, 14, 15, 22
Chapman, Colin 11, 27, 29, 30, 39, 56, 102, 123, 137, 149, 152, 156, 172, 185
Chapman, Hazel 14, 56
Cheddar Gorge 205, 207-211
Cherfas, Teresa 4, 186, 187, 189
Chobham (Longcross) test track 73, 134
Chrysler Corporation 15, 28, 160
Ciba Polymers 82
Citroën
 Dyane 58
 GS 58
 SM Maserati 222
Clark, Jim 15, 108, 109, 137, 149, 153, 184-186
Clarke, Alan 73, 124, 125, 130
Classic Team Lotus 22, 42, 137, 139, 152, 153
CLF exhausts 104
Coelho, Rui 4
Connolly hides 173
Continental Safety Engineering 72, 75
Continental Tyres 76
Cooper, John 33, 72, 73
Costin, Frank 56
Coventry-Climax 27
Crijns, Steve 4, 23, 36, 39, 62, 228
Cromer, Norfolk 120
Crowley, Mark 92, 125
Cunningham, Denis 222
Cushing, Tracey 175, 176

Da Zevio, Altichiero 183
Dallara 108
Dance, Bob 22
Dann, Katie 16, 20
D'Aste, Stefano 108, 111, 184
Daytona race circuit 106, 107
De Angelis, Elio 185
De Tomaso cars 27
Delahaye 135M 17
DeLorean DMC-12 28
Denham, Alex 4, 213
Design Museum 54, 55, 56
DFDS Ferries 186, 191
Dinnage, Chris 137

Dominguez, Pedro 4, 117
Donington Park Circuit 216
Donnelly, Martin 109, 113
Dowton, Morris 20
DRB-HiCom 23, 137
Drysdale, Laura 4
Dubai 24-Hours 111, 112
Dunster, Chris 31, 33, 68, 88
DuPont paint 160, 161, 164
Durrant, Daniel 228, 231

Earey, Max 4
Earl, Harley 56, 58
Eaton supercharger 107
Edelbrock supercharger 8, 12, 101, 117, 119
Eibach suspension 92, 110, 125, 198
Ennstal classic rally 153
Évora, Portugal 3, 15, 16, 60, 121, 144
Ewen, Stephanie 4, 205, 226
ExCel Arena 11

Farrow, Phil 159-165
Fenning, John 189
Ferrari 56, 58, 180, 189
 330 P4 40
 430 Scuderia 40, 84
 Dino 308 GT4 18, 19, 40
 Mondial 18, 40, 58
 Roma 10
Fiat
 Cinquecento 58
Fisher, Karl 128
Fittipaldi, Emerson 184
Florance, Alastair 4, 211, 216
Ford Motor Company 27, 28, 152, 160
 Cortina 58
 Galaxy 191
 GT40 40
 Mustang 40
 P68/F3L 233
Forgialluminio wishbones 126
Forsyth, Jenny 191-193
Frankfurt Show 114
Fraser, Antony 4, 127, 141, 205, 208
Fruin, Richard 155
Fuji Speedway 111

Gales, Jean-Marc 8, 22, 23, 137
Gandini, Marcello 18, 40
Gaydon Museum 137
Geely Auto Group 8, 10, 23, 24, 137
General Motors (GM) 11, 27-29, 31, 56, 58, 69, 72, 73, 98, 160
 Astra 73
 GM Speedster VX-220 34, 66, 72, 77, 80, 84
Geneva 190, 191, 193-195, 199
Geneva Show 22, 118
Gerber, Adrian 88
Giudici, Gianni 111
Giugiaro, Giorgetto 18, 40, 61
Gold Leaf Team Lotus 56
 Type 72 1, 18, 56
Goodwood Festival of Speed 85, 112, 113, 115, 148, 150, 152, 153, 225, 228, 230
Goodwood Revival 148, 149, 151, 222
Goodyear tyres 31
Grand Routière 17
Gurney, Dan 15

Hales, Mark 222
Hampton, Laura 4, 14
Hancock, Olly 111
Hangar111 108
Harrop supercharger 8, 12, 107, 108, 117
Hatt, Barney 46
Hayward, Simon 96, 101, 102, 104, 107
Haywood, David 75
Hethel test track 136, 137, 142, 144, 156, 223, 231
Heydon Hall 9, 37
Hill, Graham 149, 184
Hill, Richard 52, 53
Hofmann's 108
Hope, Brian 74
Hopper, Kerry 33
Houghton Hall 122
Hunt, Kate 4, 7
Hunter, Pete 68, 173-175
Hydro Aluminium 65, 80, 82
Hyundai 28

Ice Hotel, Lapland 70
Idiada test track 96, 104, 128-130, 132, 154
Indianapolis 106, 107, 153
 LOG28 14, 15, 29
Ireland, Innes 186
Isle of Man 216-219, 221, 222
 Manx Museum 221
 TT Course 217-221
 TT Motor Museum 222
Isuzu 27, 28

Jaguar (J-LR) 36, 54, 114
 E-type 222
 F-type 10, 17
Jaye, Andy 225
Jeffers, Alan 73
JetAlliance Racing 111, 112
Jensen cars 17, 27
Johansson, Ingemar 73, 88, 100, 124, 125
Jones, Dave 128
Jones, Matthew 16
Joyce, Richard 130
JPS (John Player Special) Type 72 56, 225

Kershaw, Gavan 4, 85, 109, 113, 126, 127, 135-137, 225
Ketteringham Hall 119, 172
Kia 28
Kimberley, Mike 4, 6, 10, 11, 16, 22, 27-31, 46, 47, 84
Kirwan-Taylor, Peter 56
Koenigsegg 40
Komo-Tec supercharger 108

Laguna Seca circuit 10, 15, 224
Lake Garda 180, 181, 183, 185
Lamborghini 29, 58
 Murciélago 40
 Uracco 18, 40
Lancia 17, 58
Land Rover 137
Larratt, Rachel 12
Lawton, Peter 31, 33, 68, 69
Le Mans 24-Hours 108, 111
Le Quément, Patrick 58
Leiston, Suffolk 202, 204
Lenman, Lena 162
Lincolnshire Wolds 213

Lingotto, Turin 11
Loch Lomond, Cameron House Hotel 146
London Motor Show: ExCel 51, 62, 63, 145, 146
Lotus Bergamo 180
Lotus Club International 180, 191, 199
Lotus Driving Academy 138-140
Lotus Engineering 28-30
Lotus Lightweight Structures 11, 65, 80, 81, 88, 167
Lotus models:
  2-Eleven 14, 15, 29, 61, 109, 141, 180, 229
  3-Eleven 61
  340R 61, 109, 141, 232
  Carlton/Omega 28
  Cortina 188
  Eclat 28
  Elan 11, 15, 30, 56, 57, 68, 69, 77, 101, 172, 188, 206
  Eleven 55, 56, 184
  Elise 6, 7, 10-12, 14-17, 22-24, 26, 27, 29-31, 34, 39, 40, 43, 44, 46-48, 52, 53, 61, 62, 65, 66, 68, 73, 77, 80-86, 90-92, 101, 102, 106, 108, 109, 122, 126, 136, 139-141, 144, 155, 159, 161, 162, 170, 174, 176, 205, 210, 231-233
  Elite 11, 15, 27, 28, 55-57, 68, 206, 232
  Emira 4, 8, 10, 61, 84, 91, 125, 138, 144, 152, 153, 176, 177, 223-233
  Esprit 8, 10, 11, 14, 15, 26-28, 30, 34, 61, 65, 69, 70, 77, 91, 96, 101, 125, 131, 132, 134, 137, 141, 143, 158, 161, 172, 229, 232
  Evija 61, 84, 85, 91, 152, 225, 227-230
  Evora 4, 6-8, 10-14, 16-19, 21-32, 34, 39, 40, 42, 43-46, 48-53, 54, 57-70, 72-81, 83-108, 112, 123-132, 134, 136-147, 155-157, 159-178, 180-204, 228, 231, 233
  Evora 400 9, 12, 13, 23, 36, 117, 118, 228
  Evora 414E Hybrid 114, 115
  Evora '50th Anniversary' 119, 120, 121
  Evora Hybrid 8
  Evora IPS auto 76, 97-100, 102, 118, 171
  Evora GT 8, 15, 121, 122
  Evora GT430/Sport 50, 52, 118-121, 229
  Evora GTE 114
  Evora Cup GT4 8, 66, 67, 108-111, 234
  Evora GTS Enduro 8, 112, 113, 114
  Evora GT410/ Sport 1, 9, 10, 18, 53, 54, 70, 116, 118, 121, 151, 175, 205-222
  Evora 'Naomi for Haiti' 117
  Evora S 8, 12, 13, 137, 139
  Evora Sport 410 118
  Evora Sports Racer 8, 115
  Europa 13-15, 27, 29, 30, 34, 40, 46, 47, 52, 61, 69, 70, 71, 80, 98, 159, 176, 188, 232
  Excel 30, 101, 172
  Exige 10, 13-15, 22, 27, 29, 30, 34, 40, 44, 52, 61, 68, 69, 86, 87, 102, 107, 136, 137, 139, 142, 158, 159, 161, 162, 174, 180, 189, 205, 209, 210, 216, 228, 229, 233
  M250 concept car 26, 40, 42, 43
  Seven 27, 81, 90, 188
  Talbot Sunbeam 28
  Type 12 184
  Type 16 81, 184
  Type 18 184, 186
  Type 21 184
  Type 23 188
  Type 25/33 56, 81, 184
  Type 30 42
  Type 38 152, 153
  Type 47 188, 232, 233
  Type 49 56, 137, 151, 184, 185
  Type 56B 151
  Type 62 232, 233
  Type 72 56, 152, 184, 225
  Type 78 184
  Type 87 81, 185
  Type 94T 185
  Type 97T 185
  Type 99T 185
  Type 115 GT1 44

Mantegna, Andrea 184
March, Lord Charles 4, 148-153
Marcos 159
Marrakech 46
Marler, Dave 74, 88-90
Martin, Guy 217, 221
Maserati 180
  Bora 18
  GranTurismo 10
Matthews, Colin 177
McLaren 36, 62, 84
McQueen, Alistair 138, 139
Merak 18, 40
Mercedes-Benz 17, 130, 143, 204, 225
MG Midget 58
Michelin tyres 53, 70, 211
Miles, John 123, 232, 233
Millbrook Proving Ground 29, 104
Mille Miglia 153, 184
Model Cars magazine 56
Moir, Glen 186
Monkey Wrench Racing 108
Montserrat 154
Monza Circuit 180, 184, 185
Morgan 27, 66, 204
  Aero 8 46
Morris
  Minor 58
Morzine 191-194
Moss, Sir Stirling 149, 186
Mowlem, Johnny 111, 112
Muir, Brian 42, 232, 233
Muirhead & Son, Andrew 174
Munday, Guy 4, 138, 139
Murphy, Chris 42

Nardo test track 96, 104, 107
Nissan 350Z 33
Norwich Cathedral Close 24
Nürburgring 24-Hours 112
Nürburgring Nordschleife 73, 123, 127-130, 143

Ogilvie, Martin 91
Öhlins coil-over dampers 112
Oliver, Jackie 185, 232
Orford, Suffolk 199, 203

Pagani Zonda 86
Paint process 159-165
Palladio, Andrea 181
Parker, Caroline 180, 191
Parnell, Jason 4, 70, 141, 180, 191, 194, 198, 199
Peck, Daniel 128
Peel, John 204
Peterson, Ronnie 184, 185
Phillippe, Maurice 56
Pinheiro, André Castro 186, 188
Pininfarina 18, 40
Pirelli tyres 11, 70, 110, 143, 195, 196, 198
Pisanello, Antonio 184
Pleavin, Andy 4, 22
Pontiac GTO 58
Popham, Phil 8
Porsche 17, 24, 28, 52, 126, 177, 184, 216, 233
  911 10, 18, 19, 24, 33, 46, 84, 107, 111, 187, 204, 210, 211, 214
  917 216
  924 Carrera GTS 153
  964 19
  991 Turbo 216
  992 Carrera 214
  997 GT3 127
  Boxster 25, 69, 84, 122, 128, 214
  Cayenne 24
  Cayman 10, 46, 84, 128, 129
  Panamera 134
Portmeirion 109
Porto 186-188
Porto Historic GP 186, 189
Portugal 186-189, 211
Portuguese Grand Prix 186
Portofino 87
ProFleet platform 226, 227
Project Eagle 8, 14, 15, 18, 31, 72, 84, 102, 103, 131
Proton 8, 11, 16, 22, 29, 75, 137
Pye, Martin 44

Qingfeng, Feng 8, 23

Rabenjamina, Harvey 229
Rackham, Richard 4, 11, 22, 31, 33, 82-88, 91, 93
Radford 62-2 232, 233
Rallye Monte-Carlo Historique 195
Recaro seats 17, 77, 174, 175, 208, 211
Redaelli wishbones 126
Reims race circuit 106
Renault 57
  4 58
  Mégane Scenic 55, 58
Rice, Anneka 122
Ricketts, Malcolm 205, 206
Rimstock wheels 53, 54, 110
Rindt, Jochen 184
Röhrl, Walter 126-129
Rolling road 176, 177
Rolls-Royce 185
Rosmolen, Joris 186
Rossiter, James 112, 113
Rover K-series engine 102
RSNürburg: Ron Simons 128, 129
Ruf Automobiles
  eRuf 104
  Greenster 104
  RT12 216

Saab 99 Turbo 58
Santander, Spain 186, 187, 189, 211
Savin, Rob 15, 16, 19, 23, 26, 27, 31, 33
Sears, Jack 122
Secker, Sandra 174
Senna, Ayrton 185
Serre Chevalier stadium 195-199
Sex, Drink, Fast Cars book 55
Shute, Tony 4, 16, 18, 19, 23, 24, 26, 27, 31, 33, 85
Sidiras, Kostas 4
Siemens 102
Silva, Ramiro Santos 188

Smart car 56, 58, 59
Snape Maltings 202
Snetterton race circuit 42, 108, 109, 113
Snow-Cat 28
SOTIRA panels 33, 34, 35, 65, 157, 159-161
Southwold, Suffolk 200, 201
Spa-Francorchamps Circuit 22
  Six Hours 222
Sport Auto magazine 128
Stanbridge Motorsport 66, 110
Stanbury, Noel 188
Stanley, Helen 225
Statham, Jon 229
Stelvio Pass 123, 130, 140
Stewart, Jackie 128, 149, 184
Surtees, Sir John 186

Tankard, Dave 73-76
Tata Motors 29
Team Dynamics wheels 54
Team Lotus 18, 56, 184-186, 188, 232
Tecalemit-Jackson fuel injection 232
Terry, Len 56
Tesla Roadster 29, 34, 58, 80, 84, 159, 223
Thomas, Neil 137
Thomson, Julian 61, 62
Tipler, Alfie 14, 19, 187, 211
Tipler, Jules 4, 205
Tipler, Zoë 19, 187, 210, 211
Top Gear 73
Toyota engines, transmissions 8, 10, 12, 13, 26-30, 69, 86, 92-103, 114, 129, 144, 168, 211, 212, 224, 227, 233
  Camry 93
  Land Cruiser 93, 94
  Lexus 93
  Prius 58
  RAV4 103
Trelleborg wishbones 92, 125
Trim shop 172-177
Triumph Motorcycles 54
TVR 159
Type approval 74, 75

Valpolicella 181
VARI molding system 28, 158
Vauxhall Motors 31
Verdi, Giuseppe 182
Verducci, Sonja 4, 234
Verona 180, 182-184
Volkswagen
  Corrado 58
Von Trips, Wolfgang 184

Waide, Martin 232
Wainwright, Peter 31, 33, 68
Warnes, Bernie 128
Webdale, James 128
Williams, Steve 125
Willment Engineering 42
Wind tunnel 39
Windle, Matt 4, 153, 223-225
Wobbly-Web wheels 55

XTRAC gearbox 112

Yokohama tyres 11, 70, 129, 141, 143, 191, 196

Zandvoort Circuit 233